Baofeng Radio
Bible

A Comprehensive Guide to Empower Your Communication,
Explore Frequencies, and Unleash the Full
Potential of Your Baofeng Radio

Gallagher Patterson

TABLE OF CONTENTS

TABLE OF CONTENTS...III

INTRODUCTION 1

CHAPTER 1 3

EVOLUTION OF BAOFENG RADIO MODELS 3

NOTABLE FEATURES AND CAPABILITIES...4

SAFETY GUIDELINES 5

SOME TYPES OF BAOFENG RADIOS..6

Baofeng UV-5R series..6

Baofeng BF-888S..8

Baofeng UV-82...9

Baofeng GT-3TP..11

Baofeng BF-F8HP..12

BAOFENG RADIOS USE CASES AND APPLICATIONS..14

UNDERSTANDING RADIO FREQUENCIES 15

VHF vs. UHF: What's the Difference?..15

But what's the difference and what does your company need?...............15

1. UHF VS VHF - WHAT'S THE DIFFERENCE? 16

VHF radio (Very High Frequency)...16

UHF radio (Ultra High Frequency)..16

Summary...16

FREQUENCY RANGES AND REGULATIONS..17

LICENSING REQUIREMENTS (IF APPLICABLE) 18

CHAPTER 2 19

BAOFENG RADIO COMPONENTS..19

RADIO BODY AND ANTENNA...19

BATTERY AND CHARGING OPTIONS 19

DISPLAY AND BUTTONS 19

The Main Display..20

CHAPTER 3 26

GETTING STARTED WITH BAOFENG RADIOS 26

UNBOXING AND INITIAL SETUP...26

Checking Included Accessories..26

Using and Installing the Antenna..26

Belt Clip 27

Chargers 28

Battery 28

How to Install a Battery 28

Charging the Battery...28
Taking out the battery 29
POWERING ON AND OFF 31
BATTERY CARE AND MAINTENANCE FOR YOUR RADIO 31
Initial Charging..31
How to Make Batteries Last Longer 31
Storage Tips 31
Dealing with Water Exposure 32
BASIC CONTROLS AND FUNCTIONS...32
Adjusting Volume and Squelch...32
SELECTING CHANNELS AND FREQUENCIES 33

CHAPTER 4 ....**35**

PROGRAMMING CHANNELS **35**

WHAT IS MEANT BY "PROGRAMMING" 35
PROGRAMMING SIMPLEX CHANNELS INTO A BAOFENG 35
PROGRAMMING REPEATERS INTO A BAOFENG...37
Set the direction of the offset..37
Then set the frequency of the offset 37
Lastly, the tone..38
HOW TO DELETE A CHANNEL 39
COMPUTER PROGRAMMING 39
Attaching the Cable...39
Baofeng Software 39
CHANNEL INFORMATION WINDOW: COLUMN DEFINITIONS...40

CHAPTER 5 ....**41**

PROGRAMMING BAOFENG RADIOS WITH CHIRP SOFTWARE **41**

HOW TO BLOCK FREQUENCIES ON BAOFENG RADIO WITH CHIRP 41
UNDERSTANDING BAOFENG RADIOS AND CHIRP 41
FREQUENCY BLOCKING: WHY IT MATTERS 42
Guarding Privacy and Security: The Why of Frequency Blocking 42
What Are the Dangers of Open Frequencies?...43
GETTING STARTED WITH CHIRP: YOUR PATH TO RADIO PROGRAMMING 43
Step-by-Step Guide to CHIRP Setup..44
THE CRUCIAL ROLE OF A COMPATIBLE PROGRAMMING CABLE 46
CONNECTING BAOFENG RADIO TO CHIRP 47
Walking through the Connection Process 47
TROUBLESHOOTING TIPS FOR COMMON CONNECTION ISSUES..48
BACKING UP RADIO CONFIGURATION 48
Backing up Radio Configuration: Safeguarding Your Radio's Digital DNA...............48
The Importance of Creating a Backup...49
A GUIDE TO SAVING YOUR CURRENT CONFIGURATION WITH CHIRP..49

CHAPTER 6 51

IDENTIFYING FREQUENCIES TO BLOCK 51

IDENTIFYING FREQUENCIES TO BLOCK: NAVIGATING THE SPECTRUM 51
    Researching and Identifying Frequencies to Block 51
    Legal Considerations When Blocking Frequencies 52
PROGRAMMING BLOCKED FREQUENCIES 52
DETAILED INSTRUCTIONS ON ADDING FREQUENCIES TO CHIRP FOR BLOCKING 53
TESTING AND VERIFYING FREQUENCY BLOCKING 54
    Explaining the Process of Testing for Frequency Blocking 54
    Troubleshooting Issues during Testing 54
CREATING AND SAVING NEW CONFIGURATION 55
    Creating and Saving New Configuration: Safeguarding Your Communication 55
    Guide to Saving the Newly Programmed Configuration 55
    Why keeping track of the configuration is a good idea 56
ADDITIONAL TIPS AND BEST PRACTICES FOR OPTIMAL BAOFENG RADIO PERFORMANCE 56
    Tips for Optimizing Baofeng Radio Performance 57
MAINTAINING FREQUENCY BLOCK SETTINGS 57
HOW TO PROGRAM BAOFENG UV-82 58
HOW TO PROGRAM BAOFENG UV-82 ON MAC – READ THE RADIO 59
RADIO TO RADIO CLONING 61
AUTOMATIC NUMBER IDENTIFICATION (ANI) 62
    Procedure: Enabling/Disabling/Configuring ANI Settings 63
SAVING AND NAMING CHANNELS 63
POWER-ON MESSAGE 65

CHAPTER 7 67

RADIO ETIQUETTE AND COMMUNICATION PROTOCOLS 67

UNDERSTANDING RADIO LANGUAGE AND TERMINOLOGY 67
HOW TO USE CTCSS AND DCS TONES CORRECTLY 68
GROUP COMMUNICATION STRATEGIES 70
    Setting Up and Managing Radio Groups 70
SELECTIVE CALLING 71
CONDUCTING GROUP CALLS AND ANNOUNCEMENTS 72
    Calling a User Group 72

CHAPTER 8 73

CTCSS 73

EMERGENCY COMMUNICATION PROCEDURES 74
    How to Enter a Basic Emergency Frequency: 74
    Reset/ "Zero Out" The Radio 74
    Select your preferred language 74
USE THE UV-5R AS AN FM RADIO 75

Enter, save, and use an emergency frequency..75

Delete a saved frequency/channel..76

SEARCH FOR ACTIVE FREQUENCIES AND TRANSMISSIONS...77

CHANGE THE RADIO'S OPERATING BAND (VHF OR UHF)..77

CTCSS AND DCS ("PRIVATE LINE" OR PL COMMUNICATIONS) 78

How to Program CTCSS and DCS to a Frequency/Channel..78

CHAPTER 9 80

INTERFERENCE AVOIDANCE 80

IDENTIFYING AND RESOLVING INTERFERENCE 80

MONITORING FOR LOCAL REGULATIONS 80

Monitoring FRS/GMRS Traffic with a UV-5R..80

One last thing and this is a big one, before we start writing. 81

You might be wondering why I didn't tell you to add any tones. 84

Per the FCC 84

SCANNING 85

How to Change the Scanner Mode 85

TONE SCANNING 86

Procedure for Tone Scanning 86

DUAL WATCH / DUAL RECEPTION 87

Functions Disabled in Dual Watch Mode 87

Procedure: Enabling or Disabling Dual Watch Mode:...87

Procedure: Locking the Dual Watch Transmit Channel 88

CHAPTER 10 89

ADVANCED FEATURES AND FUNCTIONS 89

DUAL-BAND OPERATIONS..89

EXPLORING DUAL-BAND CAPABILITIES...89

UTILIZING CROSS-BAND REPEAT FUNCTION..90

Cross Band Mini-Repeater...90

Requirements: For this setup, you'll need 91

Connection 91

ADDED CONSIDERATIONS FOR FIELD OPERATION 92

Power Conservation 92

SEPARATION AND ISOLATION 92

Frequency Selection 92

DUPLEXERS FOR SINGLE ANTENNA USE...92

Cross-Band Operation (VHF/UHF) using a Single Antenna 92

Operation on the Same Band 92

Note on Duplexer Usage 93

DIGITAL MODES AND ENCRYPTION 93

Understanding Digital Modes (DMR, D-Star, etc.) 93

IMPLEMENTING ENCRYPTION FOR SECURE COMMUNICATION 94

GPS AND APRS INTEGRATION.. 95

ENABLING AND USING GPS FEATURES 96

INTRODUCTION TO AUTOMATIC PACKET REPORTING SYSTEM (APRS) 97

    Key components of APRS include 97

CHAPTER 11 99

LEGAL AND REGULATORY CONSIDERATIONS 99

RADIO LICENSING 99

    Getting a GMRS License 99

RADIO FREQUENCY (RF) EXPOSURE LIMITS 102

CHAPTER 12 104

RADIO ACCESSORIES AND ADD-ONS 104

ANTENNA UPGRADES 104

    Why Should You Invest in Baofeng Antenna? 104

TOP LIST OF BEST ANTENNA FOR TOP 10 MODEL BAOFENG RADIO 105

1. AUTHENTIC GENUINE NAGOYA NA-771 VHF/UHF ANTENNA 105

KEY FEATURES 106

2 PACK-29 INCHES FOLDABLE/TACTICAL RADIO ANTENNA... 106

    Key Features 107

3. DUAL BAND 136-174 MHz&400-520 MHz ANTENNA FOR BAOFENG UV-82 UV-5R......... 107

4. 42.5-INCH LENGTH ABBREE SMA-FEMALE DUAL BAND FOLDABLE CS TACTICAL ANTENNA 108

    Key Features 109

5. BAOFENG BF-888S 10 X ORIGINAL ANTENNA FOR BAOFENG BF-888S 109

    Key Features 109

6. BINGFU DUAL BAND VHF UHF HAM RADIO ANTENNA 110

    Key Features 110

7. BAOFENG MAGNETIC CAR VEHICLE MOUNTED ANTENNA 111

    Key Features 111

8. BAOFENG SRH805S SMA-F DUAL BAND ANTENNA 111

    Key Features 112

9. WALKIE TALKIE ANTENNA 15.6-INCH WHIP DUAL BAND UV 112

    Key Features 113

10. HYS SMA-FEMALE HANDHELD DUAL BAND ANTENNA 113

    Key Features 113

THINGS TO CONSIDER BEFORE BUYING THE BEST ANTENNA FOR BAOFENG..................... 114

    Efficiency 114

    Antenna Compatibility 114

    Durability 114

    Size 114

CHAPTER 13 116

UNDERSTANDING ANTENNA GAIN AND POLARIZATION 116

UNDERSTANDING ANTENNA GAIN 116
    Types of Antenna Gain 117
    Markdown Table 118
FACTORS AFFECTING ANTENNA GAIN 119
MEASURING ANTENNA GAIN.. 120
HOW TO PICK THE BEST ANTENNA GAIN... 121
OPTIMIZING ANTENNA GAIN FOR BETTER PERFORMANCE 122
UNDERSTANDING ANTENNA POLARIZATION.. 123
    Why is antenna polarization important for amateur radio?.. 125
    Different types of antenna polarization 126
FACTORS AFFECTING ANTENNA POLARIZATION.. 127
HOW TO DETERMINE THE POLARIZATION OF AN ANTENNA 128
PRACTICAL APPLICATIONS OF ANTENNA POLARIZATION IN AMATEUR RADIO 130

CHAPTER 14 132

CHOOSING THE RIGHT PROTECTIVE GEAR.. 132

ACCESSIBILITY AND COMFORT CONSIDERATIONS.. 133
HARDWARE MODIFICATIONS 135

CHAPTER 15 137

INTERNATIONAL RADIO REGULATIONS.. 137

GLOBAL FREQUENCY ALLOCATIONS.. 137
UNDERSTANDING ITU RADIO REGULATIONS.. 138
BACKGROUND AND PURPOSE OF ITU RADIO REGULATIONS 138
    Spectrum Allocation and Management 138
    International Coordination.. 138
    Framework for Innovation and Development 139
    Key Components of ITU Radio Regulations 139
    Significance and Impact 140
CHALLENGES AND EVOLVING LANDSCAPE 140
    Spectrum Scarcity.. 140
    Emerging Technologies 140
    Regulatory Adaptation.. 141
HARMONIZING BAOFENG RADIOS FOR INTERNATIONAL USE 141
FREQUENCY BANDS AND COMPLIANCE 141
    Regional Frequency Allocation... 141
    Coverage of Frequency Range 141
TRANSMIT POWER LEVELS 142
    Power Output Limitations 142
TYPE APPROVAL AND CERTIFICATION.. 142
    Regulatory Compliance and Certification 142
COMPLIANCE PROGRAMMING AND MODIFICATION 142
    Programming and Locking 142

User Education and Awareness..142

    Regulatory Awareness 142

CHAPTER 16 ....**143**

MAINTENANCE AND TROUBLESHOOTING **143**

Care and Maintenance Tips 143

    Cleaning and Handling the Radio..143

    Cleaning 143

    Handling..144

Battery Care and Replacement 144

    Battery Care 144

    Battery Replacement 145

Troubleshooting Common Issues 145

    Baofeng UV-82 Troubleshooting..145

Troubleshooting Common Issues on Baofeng Radio Programming 147

Seven Steps to Avoid Most Baofeng Radio Problems 147

Frequently Asked Questions 151

    What is a Baofeng?..151

    Can I use it for amateur radio? 151

    Can I use it for PMR 446MHz? 152

    Can I use it for Ofcom Business Licence?..152

    Are there any frequencies I can transmit without a license? 153

    Which Baofeng radio is the best? 153

    Does Baofeng receive frequencies from aircraft? 153

    How much do Baofeng radios cost?..153

    Can I make my Baofeng radio's range better?..153

    Does Baofeng radio come with a warranty?..153

    How many Watts do Baofeng radios have?..153

    Can I use the radio for general communication with family? 154

    Can I use the radio as a scanner?..154

    How far does the Baofeng UV 82 work? 154

Conclusion 154

INDEX **155**

INTRODUCTION

Welcome to the exciting world of Baofeng radios, where staying connected is as easy as pushing a button! This guide is your go-to handbook for understanding and using Baofeng radios in a way that's simple and stress-free. It's so important to be able to talk to your family and friends in a situation that it almost goes without saying. It's important to plan what to do if your phone or internet goes out. It's not hard to think about since this happens all the time. Still, not many people have a real plan for how to communicate with their neighbors in case of an emergency. For less than $30, you can get a gadget that lets you connect with people in your area, hear messages about bad weather, get minute-by-minute updates on what first responders are doing, and even call for help if you need to. The Baofeng Radio is a cheap first-time radio that every family that is planning to be ready for anything should have. Have you ever thought about how radio works? Baofeng is here to clear everything up. Imagine a world where using your favorite device is all it takes to talk to friends, plan outdoor activities, and be ready for situations. This Baofeng radio is like a smart walkie-talkie. It can talk on different channels, switch between city and country frequencies, and even change its power to save battery life, which are all cool features. Do not worry about having a tech background; we will help you with everything.

Baofeng radios are like superheroes when it comes to talking on the phone. No matter what you need, they can talk over short lengths or a large area. These radios are made to make you feel like a communication superhero, no matter how experienced you are or how new you are to the game. This guide will walk you through each step like a friend. We'll talk about everything in simple terms and give you some secret tips so that you can use your Baofeng radio with confidence no matter what. So, buckle up for an adventure in communication made easy. Let's get started on your journey with Baofeng radios – where simplicity meets connectivity!

CHAPTER 1
EVOLUTION OF BAOFENG RADIO MODELS

The history of Baofeng radios is like an interesting journey through time. These radios have changed a lot over the years to meet the needs and wants of all kinds of users. Since Baofeng was new to the market, their first radios were very basic and very cheap. For many, these radios, like the UV-5R, were like magic ways to talk to each other. They only had a few functions, but a lot of people could use them to easily talk to each other from far away. Over time, Baofeng made their radios better by adding cool new features. They made radios with longer battery lives so that people could use them for longer. The number of channels on these new types increased the number of options for conversing with various individuals. That wasn't all Baofeng did. They began making radios that could be used for different things. Some were made to be strong so they could handle rough conditions. This makes them great for situations or trips outside. Other types were made to have more power, which would let people talk over longer distances. There was something for everyone with these different kinds of radios. As technology got better, Baofeng made their radios work with digital technology. Brand-new models, like the DM-5R, had better sound quality and more protection options because of this. It was like switching from an old phone to a new smartphone—you could tell and enjoy the change in sound quality and speed.

Baofeng also changed the way their radios felt and looked. They added bigger screens and easier-to-press buttons to make them easier to hold and use. These changes were made so that the radios would be easier to use and more fun. In later models, Baofeng radios could even connect to other devices. They could connect to headphones or phones, which made it easier and more flexible to talk to people. These new ideas let people talk to each other in different ways, which made the radio more useful. When we think about the future, it looks good for Baofeng radios. Their technology could get even better, which would make them smarter and last longer. Soon, there may be new features like voice prompts or smooth interaction with other devices that will make the user experience even better. In a nutshell, Baofeng radios have changed over time in an exciting way that keeps getting better. They started as easy ways to talk to each other and have grown into complex tools that can be used for many things. Better conversation, greater ease of use, and more options for keeping linked in different scenarios have all been added to Baofeng radios with each new version.

Notable Features and Capabilities

Over the years, Baofeng radios have changed a lot. They now have a lot of unique features and functions that make them stand out among small transceivers.

Here are some things that make Baofeng radio models stand out:

❖ **Reasonable prices**: Baofeng radios are known for being very affordable. Because they are affordable and have a lot of features, they can be used by a lot of people.

❖ **Dual-Band Operation**: Many Baofeng radios are compatible with both VHF (Very High Frequency) and UHF (Ultra High Frequency) bands, giving users flexible contact options across various frequencies.

❖ **Programming Flexibility**: These radios let users program frequencies and channels by hand, so they can change the settings to suit their wants and tastes.

❖ **Compact and Portable Design**: Baofeng radios are known for being easy to hold in one hand and being light. This makes them great for outdoor activities, emergencies, and work.

❖ **Longer Battery Life**: Some models have longer battery lives, which means you can use them for longer periods, which is especially helpful when you need to talk for a long time.

❖ **Broader Channel Capabilities**: Most Baofeng radios have a lot of channels, so users can talk on a lot of different frequencies and be flexible in a lot of different settings.

❖ **Support for Digital Mobile Radio (DMR):** As digital technology has improved, some Baofeng models now support DMR, which provides better sound quality, more protection features, and better use of airwaves.

❖ **Built-in flashlight features**: Some types come with built-in flashlight features that make the radios more useful in low-light situations or accidents.

❖ **The VOX (Voice-Activated Transmission)**: This feature on Baofeng radios lets users send messages without using their hands or hitting the push-to-talk button.

❖ **Fits a Variety of Accessories**: These radios work with a variety of accessories, including antennas, earpieces, mics, and batteries, which makes them more useful and functional.

- ❖ **User-Friendly Interface**: Many Baofeng models have simple, easy-to-understand screens with buttons that make travel and operation simple, even for newbies.
- ❖ **Robust Construction**: Baofeng radios are known for being long-lasting and well-built, able to handle rough use and a wide range of weather conditions.
- ❖ **Actual-time Scanning and Monitoring**: Some Baofeng radios may let users move between different channels and frequencies to stay up to date on current conversations.

Together, these useful features and functions make Baofeng radios reliable, flexible, and able to meet the contact needs of a wide range of people, from amateur radio fans to professionals working in many fields.

Safety Guidelines

- **Qualified Service**: This gadget should only be serviced by trained techs. Do not mess with or change the radio by yourself.
- **Use Approved Accessories**: To avoid any problems, only use batteries and chargers that have been approved by BAOFENG.
- **Damaged Antenna Warning**: Do not use a radio that has a broken antenna. A small burn could happen if they touch.
- **Safety in Dangerous Areas**: Turn off the radio before going into a room with bombs or burning materials.
- **Safety when charging**: Never charge the battery in a place where some bombs or things can catch fire.
- **Avoid Interference**: To avoid interference or problems, turn off the radio in places where it says not to use it.
- **Air Travel Rules**: This says that you must always turn off the radio before getting on a plane. When you use a radio on a plane, follow the rules set by the company.
- **Safety in Explosive Areas**: To stay safe, turn off the radio before going into blast zones.
- **Be careful with airbags**: If your car has airbags, don't put the radio where an airbag could go off.
- **Protect from Heat and Sun**: Don't leave the radio out in full sunlight for long periods or put it near heat sources.

- **Safe transmission**: Hold the radio straight up and keep the microphone 3 to 4 cm away from your lips. During communication, make sure the receiver is at least 2.5 cm away from your body.

These simple safety tips will help you use and handle your radio safely and stay away from possible dangers.

Some types of Baofeng Radios

Baofeng UV-5R series

The Baofeng UV-5R series is a group of useful mobile dual-band transceivers that are very popular with amateur radio operators, first responders, survivalists, and other people who use radios for contact. These radios are known for being affordable, small, and flexible. They have a huge number of features that make them useful for a wide range of communication needs.

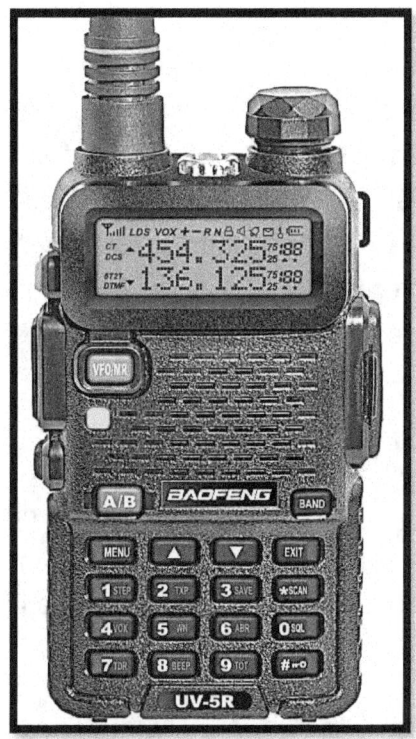

In terms of design, the UV-5R series has a tough but soft feel to it, with a strong build that is easy to hold. It is very movable because of its small size, and it fits comfortably in your hand for long amounts of time. The LCD screen clearly shows important details like frequency, channel, and power state on the easy-to-use interface. The buttons and knobs are set up in a way that makes sense, making it easy to get to features and settings. One thing that makes the UV-5R series stand out is that it can work on both the VHF (Very High Frequency) and UHF (Ultra High Frequency) bands. This lets people talk on a wide range of frequencies, giving them more options and services in different situations. The dual-band feature makes it possible to work in a variety of settings and frequencies, whether you're talking to people nearby or sending data over longer distances. With broadcast powers ranging from 1 to 5 watts, these radios make it easy to talk clearly and reliably even in rough terrain or places with lots of hurdles. They also work with a lot of different frequency ranges, so you can use a lot of different channels and frequencies within the bands that have been given to you.

The UV-5R series comes equipped with a multitude of features, including but not limited to:

- **Channel Memory**: The radios can save a lot of channels so that they are easy to get to and make switching between frequencies that you use often a breeze.
- **Voice Operated Exchange (VOX):** This hands-free function lets people send messages without pressing the Push-To-Talk (PTT) button. When the device detects voice input, it starts the message-sending process.
- **Dual Watch and Dual Reception:** You can keep an eye on two channels at the same time with this feature. It makes sure that users stay up to date on multiple frequencies.
- **FM Radio Receiver**: These radios often come with an FM radio receiver built in so users can listen to their favorite FM stations.
- **Emergency Alert:** The UV-5R series usually has an emergency sound feature, which is very important for getting people's attention in an emergency.
- **Battery Life**: These radios work reliably and have a good battery life. Some types even let you switch out the batteries for longer use.

The Baofeng UV-5R series is also known for being flexible and able to work with a wide range of devices, including antennas, mics, and programming cords. Users can change and improve their conversation experience based on their own needs and tastes thanks to this freedom. Overall, the Baofeng UV-5R series comes out as a reliable, feature-packed, and cost-effective choice for people who want a strong pocket radio that can meet a wide range of communication needs. This radio is a popular choice among both amateurs and pros because it is versatile, easy to carry, and not too expensive.

Baofeng BF-888S

A basic portable radio that is popular and known for being simple, reliable, and easy to use is the Baofeng BF-888S. It's a simple model, but it has important features that make it useful for a wide range of communicating needs. **Here are some of the most important things about the Baofeng BF-888S:**

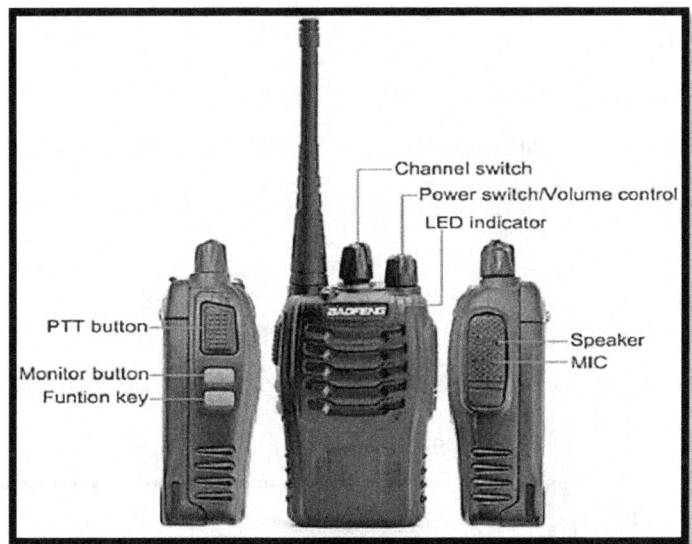

- **Frequency Range:** It works in the UHF band, with a frequency range of 400–470MHz, so it can be used for short-range conversations in cities or inside buildings.
- **16 Channels**: It has 16 customizable channels, so users can talk on more than one channel and reach different frequencies.
- **Compact Design**: The BF-888S is small and light, so it's easy to carry and perfect for people who want a pocket phone.
- **Long Battery Life:** It comes with a big battery that lets you use it for a long time, so you can be sure that your connection will work for a long time.
- **Power Output**: Has two power sets that can be chosen—high and low—so that the transfer power can be changed based on the contact range.

- **Squelch Levels**: You can change the squelch levels to reduce background noise and make conversations clearer.
- **VOX Features**: It supports Voice-Activated Transmission (VOX), which lets you use it hands-free when you use it with suitable devices.
- **Simple Operation**: The interface is clear and easy to use, so even people who have never used computers before can do it without any problems.
- **Emergency Alarm**: It has an emergency alarm button that sends out a signal in case of an emergency or to call for help.
- **Durable Build**: The build is strong and long-lasting so it can handle rough handling and everyday use in a variety of settings.

Even though the Baofeng BF-888S is a basic model, it is praised for being reliable and easy to use. This makes it ideal for businesses, security teams, casual users, and other situations where simple but effective communication within a short range is needed. It is a reliable choice for short-distance contact because it is easy to use and well-built.

Baofeng UV-82

The Baofeng UV-82 series is a well-known group of two-way radios that are mobile and work on two bands. People who like radio, amateur radio operators, and workers in many fields choose these radios because they are good value for money, can be used in many situations, and have many features.

1. **Dual-Band Capability**: The Baofeng UV-82 series' ability to use two bands is one of its best features. These radios can work on both VHF (Very High Frequency) and UHF (Ultra High Frequency) bands, which gives people more options and access to a wider range of frequencies.

2. **Frequency Range**: The UV-82 series has a large field of view, ranging from 136 to 174 MHz for VHF and 400 to 520 MHz for UHF. Because these radios cover a wide range of frequencies, users can talk on different radio bands, which make them useful for many situations.

3. **Output Power**: These radios are known for having power sets that can be changed so that users can choose between high and low power output levels. The power level can be changed, which can help save battery life or make sure that transmission works well over greater distances.

4. **LCD Screen**: The UV-82 series has an LCD screen that is clear and easy to use. The screen shows important details like the chosen station, frequency, power level, and more. This function makes the radio's settings easy to find and improves the user experience.

5. **Battery Life**: Most Baofeng UV-82 radios come with lithium-ion batteries that can be charged again and again. The battery life can change depending on how the power is set and how often it is used. However, these radios are known to have good battery life, and extra battery packs are often available for longer use.

6. **Programming and Software**: The UV-82 series can be programmed using computer software, which lets users change settings like channels, frequencies, and more. This function is especially helpful for amateur radio users who might want to set up specific rebroadcast frequencies or make their channel lists.

7. **Compact Design**: These radios are small and comfortable to hold because of their design. The sturdy construction makes them last a long time and makes them good for a wide range of outdoor activities, emergencies, and work.

8. **Voice-Operated Exchange (VOX) Functionality**: The UV-82 series has the VOX feature, which lets people use the radio without using their hands. This can be helpful in cases where operating by hand might not be possible or would be difficult.

9. **Dual Watch and Dual Reception**: The UV-82 series has dual watch and dual reception features that let users watch two channels at the same time. You can

stay up to date on events on multiple frequencies without having to switch between channels all the time.

10. **Accessories and Compatibility**: Antennas, earpieces, mics, and programming connections are just a few of the items that Baofeng makes for the UV-82 series. The radios can also often work with devices made by other companies, which makes them more flexible.

11. **Legal Compliance**: People who want to use the Baofeng UV-82 series should be aware of any area laws and license requirements. In many countries, you need a license to send signals on certain frequencies, and you must follow these rules.

12. **Community and Support**: There are a lot of people who use the Baofeng UV-82 series. Online communities, user groups, and boards let people share their experiences, work out problems, and give and receive tips and tricks for using and getting the most out of these radios.

Baofeng GT-3TP

The Baofeng GT-3TP is a strong mobile transmitter from Baofeng's well-known GT-3 series. It's an updated version with more features and more power, making it perfect for amateur radio operators, first responders, and anyone else who needs to communicate clearly in a variety of situations.

The design of the GT-3TP is similar to that of its predecessors. It is sturdy and well-thought-out so you can use it for long periods without getting tired. It is easy to use for both new and experienced radio fans because it is well-built and the buttons and settings are laid out in a way that makes sense. One thing that makes the GT-3TP stand out is that it has more motor power than normal models. This transmitter can send with a power level of up to 8 watts, which means it has a stronger signal and a wider contact range than its predecessors. This extra power makes it easier for it to keep contact clear and reliable, especially in rough landscapes or places where there is an interruption. The GT-3TP works on both the VHF and UHF bands, just like other Baofeng types. This gives users access to a wide range of frequencies for flexible contact. Its dual-band capability makes it flexible in different settings by covering a wide range of frequency bands.

Key features of the Baofeng GT-3TP include:

- ❖ **Better Power Output**: This radio's highest power output of 8 watts makes transmissions stronger and contact range longer.
- ❖ **Dual-Band Operation**: Operating on both VHF (136–174 MHz) and UHF (400–520 MHz) bands, this feature gives you the freedom to use a variety of frequencies.
- ❖ **High-Gain Antenna**: Usually comes with a high-gain antenna that makes receiving and sending signals better.
- ❖ **Large Capacity Battery**: The GT-3TP usually comes with a large-capacity battery, which lets you use it for longer and avoids having to charge it often.
- ❖ **Multiple Functions and Features**: It has many useful features, such as two watches, scanning functions, VOX support, FM radio receiving, and an emergency warning, which make it more useful in many situations.
- ❖ **Compatible with Accessories**: The GT-3TP, like other Baofeng radios, can work with a variety of accessories, such as antennas, speakers, and programming cords. This lets users adjust and improve their communication experience.

Baofeng BF-F8HP

The Baofeng BF-F8HP is a well-known portable transmitter that is known for its strong features, high power output, and flexibility. It was made to meet the needs of amateur radio fans, first responders, and other communication pros.

This model is an updated version of the Baofeng UV-5R series, which is known for having better speed and more power. The BF-F8HP keeps the standard Baofeng design, with a tough but soft build that makes it easy to hold and use. One thing that makes the BF-F8HP stand out is how much power it puts out. It can send signals at up to 8 watts, which is a lot more than most mobile transceivers in the same class. This extra power helps to boost the signal strength, extend the range, and make contact more reliable, especially in places where there are hurdles or interruptions.

Key features of the Baofeng BF-F8HP include:

- ❖ **High Power Output**: With a maximum output of 8 watts, it has better range and communication power than regular mobile radios.
- ❖ **Dual-Band Functionality**: It works on both the VHF (136–174 MHz) and UHF (400–520 MHz) bands, so you can use a lot of different frequencies for different kinds of conversations.
- ❖ **Large Battery Capacity**: Comes with a large battery that lets you use it for longer amounts of time without having to charge it often.
- ❖ **Upgraded Antenna**: It usually comes with a better antenna that makes it easier to receive and send signals.
- ❖ **Multiple Functions and Features**: The BF-F8HP has many of the same features and functions as other Baofeng models. It has a dual watch, scanning capabilities, VOX

function, FM radio reception, emergency alarm, and more, so it can meet a wide range of communication needs.

- ❖ **Backlit LCD Display**: This is a screen that shows important details like frequency, channel, and battery life, even when there isn't much light.
- ❖ **Compatibility with Accessories**: The BF-F8HP works with many different accessories, including radios, speaker mics, programming cords, and more. This lets users make their contact experience more personal and effective.

Baofeng Radios Use Cases and Applications

Because they are flexible, useful, and easy to use, Baofeng radios can be used in a wide range of situations. Here are some real-life situations in which Baofeng radios can be useful:

- ❖ **Outdoor Activities**: Baofeng radios are great for people who love being outside and can be used while climbing, camping, shooting, and other outdoor activities. They make it easier for people in a group to talk to each other over different surfaces and distances.
- ❖ **Emergency Preparedness**: Baofeng radios are important tools to have in case of a disaster. They let people talk to each other during situations or natural disasters when regular networks might not be available.
- ❖ **Recreational Use**: Radio hobbyists, fans, and amateur operators use Baofeng radios to talk to each other, try out new frequencies, and meet with other radio users.
- ❖ **Event Management:** People who plan and run events depend on these radios to make contact easy at festivals, sports games, concerts, and other public gatherings where quick and clear communication is needed.
- ❖ **Search and Rescue Operations**: Baofeng radios are very important for search and rescue tasks that first responders do because they let relief teams work together and talk to each other.
- ❖ **Construction and Security**: Baofeng radios help workers, managers, and security staff stay in touch with each other on construction sites and in security activities, which improves speed and safety.
- ❖ **Hiking and Mountaineering**: Backpackers, climbers, and mountaineers use these radios to stay in touch in rural and hilly areas, which keeps everyone safe and helps team members work together.

- ❖ **Remote Areas and remote Communication**: Baofeng radios are an important way for farmers, ranchers, and people to talk to each other in remote areas or places where there is limited cellular network coverage.
- ❖ **Volunteer Groups**: These radios are used for organized activities and contact by volunteer groups like community guards, neighborhood watch teams, and crisis aid groups.
- ❖ **Professional and Industrial Use**: Baofeng radios are used to communicate between teams, coordinate logistics, and handle operations in fields like logistics, manufacturing, and transportation.
- ❖ **Education and Training**: Radio fans, radio lovers, and people who want to be radio experts use Baofeng radios to learn about radio transmission and get licenses.
- ❖ **Exploration and Field Research**: Baofeng radios are used by scientists, researchers, and explorers to talk, share data, and work together in rural places and outdoor settings.
- ❖ **Maritime and Boating**: Some Baofeng types can be used in the water, making it easier to talk to people on ships, and boats, and while traveling.
- ❖ **Family Communication**: Baofeng radios help families and groups stay in touch while on vacation, a road trip, or an event, especially when cellular network coverage is limited.

Understanding Radio Frequencies

VHF vs. UHF: What's the Difference?

When shopping for two-way radios, one of the most important things to think about is whether you want VHF or UHF radios. Based on the size of their business, their budget, and where their team is located, each needs a different time range.

But what's the difference and what does your company need?

The main difference between these two options is in the range of frequencies that they use. But there's more...

1. UHF vs VHF - what's the difference?

VHF radio (Very High Frequency)

VHF radios use radio waves between 30 MHz and 300 MHz, which is a wider broadcast area. They don't have as many channels as other two-way radios close, which can cause problems with crowding and crosstalk. Generally, VHF radios are less expensive and have been around for much longer than UHF radios. Two-way radios with very high frequencies work best outside and over big areas. This is because their signals weaken in cities with lots of barriers, like tall buildings. Also, they work best in places with few people, so they can do their job without being hampered. On a general level, VHF radios work better in fields like farm and leisure.

UHF radio (Ultra High Frequency)

There is less range for UHF two-way radio waves because they can't go as far as VHF radio waves. But because they cover a wider frequency range, they offer better service with fewer disturbances from other users. UHF radios use up battery power faster because they work on a higher frequency. This means that workers shouldn't use them if they can't easily get to a charging point. One great thing about UHF radios is that they can get through things like concrete, steel, and wood which are in the way better in cities. Because of this, this type of radio works best indoors, even in buildings with more than one floor, in fields like healthcare, education, retail, manufacturing, and delivery. They also work well for companies that do business both inside and outside and where there are a lot of other buildings.

Summary

- Radios with lower frequencies (VHF) have longer ranges, which makes them great for working in large areas with few obstructions and outside.
- UHF radios have a higher frequency, which makes them perfect for wireless conversations that need to go through walls, buildings, concrete, and other obstacles. Because of this, UHF radios work best for talking inside, where there are things like walls that could get in the way.

Frequency Ranges and Regulations

Baofeng radios, such as the UV-5R series, the GT-3TP, and the BF-F8HP, work within particular frequency bands that are controlled by political organizations. Very High Frequency (VHF) and Ultra High Frequency (UHF) bands are the ones that these radios usually work with. But it's important to know that rules about which frequencies these radios can use may be different in different countries, and users must follow local laws and license requirements.

These frequency ranges are what most Baofeng radios can handle:

1. VHF (Very High Frequency):
   - The frequency range is from 136 MHz to 174 MHz.
   - This band is used for many things, like public safety, sea, flight, amateur radio (ham radio), and FM broadcasts.
2. UHF (Ultra High Frequency):
   - 400 MHz to 520 MHz frequency range
   - UHF is used for services like business radio, public safety radio, amateur radio, and more.

When it comes to these bands, Baofeng radios usually have access to a wide range of frequencies. This lets users talk on different channels that are set aside for specific reasons. It's important to keep in mind, though, that not all frequencies in these areas can be properly accessed without the right licensing. Different areas and countries have different rules about how to use these radios. It is often against the law to send signals on certain frequencies, especially those that are only for emergency services or government use, without the right permission or license. The Federal transmission Commission (FCC) in the US, for example, is in charge of radio transmission. People who want to use Baofeng radios must get the right FCC license to send on amateur (ham) radio frequencies. To stay out of trouble with the law, users must also follow the rules for other services or frequencies. Before using Baofeng radios, users should make sure they know the local rules and license requirements to make sure they are using the frequency bands legally and responsibly. Getting the right licenses or permissions is important if you want to stay within the law and help promote sensible radio communication.

Licensing Requirements (if applicable)

Having a thorough understanding of the regulations is essential prior to making use of a Baofeng radio in the United States. Regulatory guidelines for radios such as the Baofeng are established by the *Federal Communications Commission (FCC)* in the United States. According to them, in order to operate specific kinds of radios, including the Baofeng, you are required to get a license. You can submit an application for a specialized *General Mobile Radio Service (GMRS)* license if you want to use the Baofeng for GMRS. All members of your immediate family are covered by this license, which is valid for a period of ten years. It is necessary to get a license from the Federal Communications Commission (FCC) in order to use the Baofeng radio as an amateur radio. To get this license, you are required to first obtain an FCC Registration Number (FRN) and then pass an examination. By using a Baofeng, you are permitted to listen to amateur radio; but, in order to speak or broadcast, you are required to have a valid amateur radio license issued by Ofcom in the United Kingdom. A Baofeng radio must comply with the guidelines established by the FCC in order to be used lawfully. This indicates that it has to be properly authorized, used within the appropriate frequency ranges, and not possess an excessive amount of power.

CHAPTER 2
BAOFENG RADIO COMPONENTS

Radio Body and Antenna

People like Baofeng radios because they are cheap and can do many things. They are made of tough plastic that isn't too heavy, so they're easy to move around with you. A small screen on these radios shows what station they're on, how much battery life is left, and other settings. These radio antennas are very important because they send and receive messages. Most of the time, these radios come with a standard antenna that works just fine. But you can get a better receiver if you want your radio to work farther away. A few well-known ones are the Baofeng Magnetic Antenna 1, the Radtel Foldable Tactical Antenna, and the Nagoya NA-701C. These antennas are made to make it easier for your radio to send and receive messages.

Battery and Charging Options

Baofeng radios have many power options so users can choose the one that works best for them. Utilizing the normal charger that comes with the radio is the main and most usual way. This charger usually comes with a USB cord that lets you connect it to different power sources, like computers or wall plugs. However, there are more than just normal ways to charge Baofeng radios. They can also be charged up in other ways, such as by using sun energy, extra USB cords, or 12V cigarette plugs. Getting a multi-unit charger is something that people who want to charge their Baofeng radios more efficiently should think about. The CH-5-6 Gang Charger is a great choice because it can charge up to six single-cell lithium-ion batteries, six dual-cell lithium-ion batteries, or six six-cell nickel-metal hydride batteries at the same time. The UV-5R Series Six Way Charger is another good choice. It can charge up to six radios or batteries at the same time, which makes the process of charging multiple devices faster.

Display and Buttons

A Baofeng radio's body is usually made of tough plastic to make it last a long time while still being light enough to carry around. The body of this radio has a screen that shows important information like the radio channel, power level, and different settings.

The Main Display

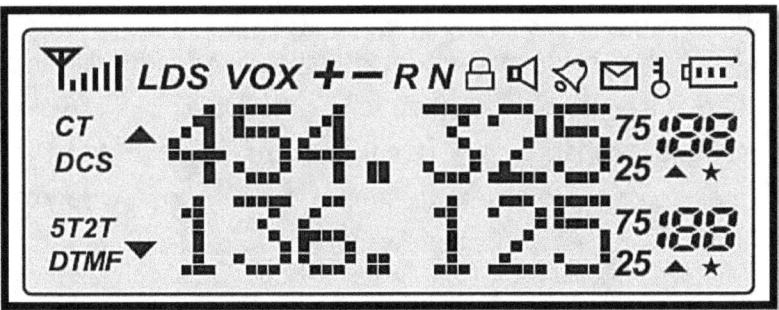

Icon	Description
188	Memory Channel
75/25	Least Significant Modifiers
CT	CTCSS enabled
DCS	DCS enabled
+ -	Frequency Offset Shift Direction if Enabled
S	Dual Watch/Dual Reception if Enabled
VOX	VOX Function if Enabled
R	Reverse Function if Enabled

N	Narrowband Enabled
(battery icon)	Battery Level Indicator
(key icon)	Keypad Lock Function if Enabled
L	Low Transmit Power Mode if Enabled
▲ ▼	Active Band or Channel
(signal icon)	Signal Strength Meter

Some buttons on the radio can be used to change settings and move through its functions:

The VFO/MR button on your Baofeng radio is very useful because it lets you easily switch between channel mode and frequency mode. In channel mode, the screen shows a full list of all the saved frequencies along with their unique call signs, like "K5QHD" in this case. These call signs are used to identify registered repeaters, which makes them easy to find when they are being used. However, frequency mode allows you to directly enter frequencies using the keyboard, providing a more individualized approach to the radio setting.

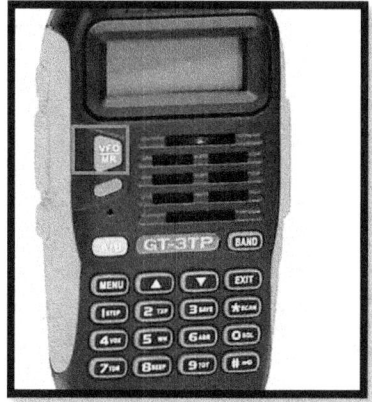

The **A/B button** is also very important because it makes switching between the top and bottom rows of channels very easy. You can easily switch between these channels by pressing the arrow keys on your keyboard.

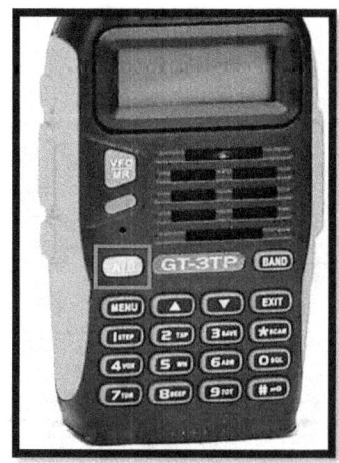

The **Band button** is very important because it lets you switch between the 2M VHF band (144–148MHz) and the 70CM UHF band (420–450MHz), which gives you access to different frequency ranges.

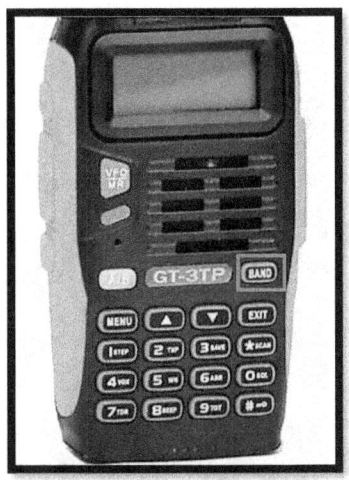

As for the **Scan feature**, it is very important for getting to know nearby repeaters and radio action. You can look through channels in a planned way by holding down the look button. This is especially helpful when you are first learning about the repeaters in your area. It is important to note that the skies may seem quiet at times, but hams usually use networks in the night, Monday through Wednesday, from 7 to 8 p.m. It is suggested that you scan your area often and at different times of the day to get to know it.

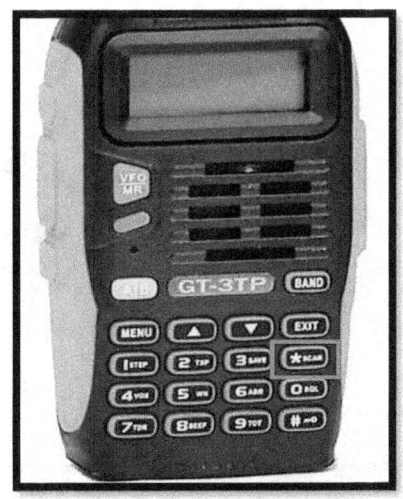

The "Call" button sets off a warning that is sent out on the frequency you are currently using. To avoid interruptions, you should never press this button while constantly looking through channels.

To communicate, you need to press the *PTT (Push-To-Talk) button*. However, you must have a ham license to use this feature. Also, you should never use the radio without an antenna connected.

Battery Level Indicator:
- When the battery level indicator shows no black bars, it means the battery is nearly empty.
- To alert you about this, the radio will periodically beep and flash the display's backlight, signaling that it's time to change the battery.

Status LED:
- The status LED has a simple design to indicate different modes of operation:
  - When receiving a signal, the LED turns green.
  - During transmission, it switches to red.
  - In standby mode, the LED remains off.

Last but not least, the MONI button on the bottom turns off the mute so that the sound levels can be checked. This makes the speaker make static noise that the squelch function normally blocks out, which helps you judge the volume.

Pound Button:
- ❖ The [POUND] button lets you quickly switch between High and Low broadcast powers when you're in channel mode.
- ❖ Keep in mind that this setting change will only last for this session and will not affect the sending power saved in the channel's memory in the future. When you switch to a different channel or working mode, the broadcast power is reset to the memory setting.

Keypad Lock:
- ❖ The Baofeng UV-82 has a keypad lock function that blocks all but the three buttons on the side of the radio.
- ❖ Hold down the [**POUND**] button for about two seconds to turn on or off the passcode lock.
- ❖ From the menu, you can also turn on automatic keypad locking, which locks the keypad after ten seconds of idleness.

Star Button:
- ❖ A quick press of the [**STAR**] button enables the reverse function.
- ❖ To start scanning while listening to broadcast FM, press the button for a short time. The screening stops when a live station is found.
- ❖ Hold down the [**STAR**] button for two seconds to make scanning go on forever.

CHAPTER 3

GETTING STARTED WITH BAOFENG RADIOS

Unboxing and Initial Setup

Checking Included Accessories

There is a wide range of extras for Baofeng radios that can make your general user experience much better.

Here is a list of some of these items that could be very useful:

Wi-Fi antennas: Most Baofeng radios come with a basic antenna that works well for everyday use, but you may want to make your radio's range better. This can be done well by upgrading to an antenna with better strength. The Nagoya NA-701C, ABBREE AR-152A, Radtel Foldable Tactical Antenna, and Baofeng Magnetic Antenna are some of the most well-known options in the world of Baofeng radio antennas.

Using and Installing the Antenna

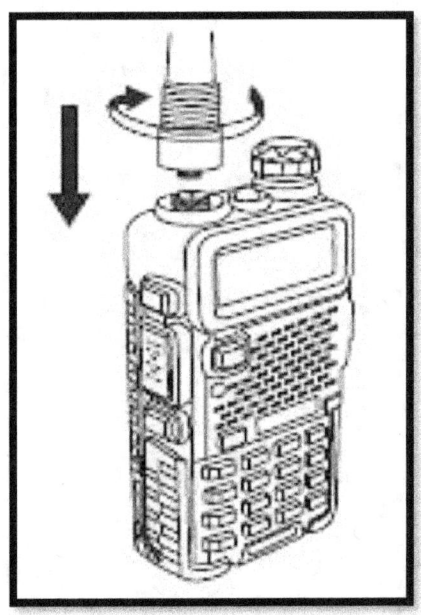

1. **Connector Alignment**: The antenna has a Female SMA Connector and the device has a Male SMA Connector. To place the antenna tightly on the device, line up these two links and twist the antenna clockwise until it stops.
2. **Proper Installation** The right way to install an antenna is to hold its base tightly and then twist it into place.
3. **External Antenna Usage**: If you use an external antenna, make sure the "SWR" (Standing Wave Ratio) is 1.5:1 or less. This helps keep the transistors in the receiver from getting damaged.
4. **Care when handling**: Don't hold the antenna in your hand or wrap it around something, as this could affect how well it works.
5. **Antenna Essential for Transmission**: Never use the gadget to send messages without connecting an antenna first. This could hurt the gadget.

Belt Clip

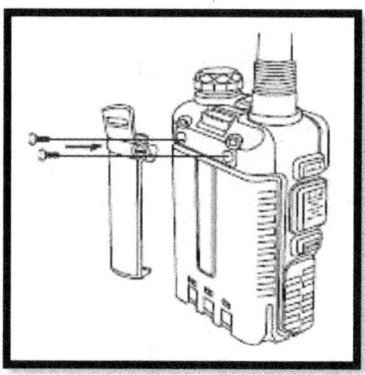

❖ **Find the screws.** There are two screws next to each other on the back of your radio, above the battery. To take them off, unscrew them.
❖ **Align with Belt Clip Holes**: Line up these screws so they go through the holes on the belt clip.
❖ **Secure the Belt Clip**: Put the belt clip against the back of the battery cover so that the holes in the clip line up with the pins on the radio body. After that, put the screws back into the radio to make sure the belt clip stays in place.
❖ **Don't Use Glue**: When you put the screws on the belt clip, don't use glue. Putting glue on the battery could hurt its case. It's important to use the screws that come with the belt clip to keep it in place.

Chargers

Baofeng radios can be charged in a variety of ways. The usual way is to use the charger that came with the radio, which is usually a USB cord that connects to a power source like a computer or wall adapter. But these radios are flexible, and they can be charged in different ways, such as with sun power, USB cords, or 12V cigarette plugs. Buying a multi-unit charger can be helpful if you want to charge more than one device at once.

- ❖ For example, the CH-5-6 Gang Charger can charge up to six single-cell lithium-ion batteries, six dual-cell lithium-ion batteries, or six nickel-metal hydride batteries at the same time. The UV-5R Series Six Way Charger, on the other hand, can charge up to six radios or batteries at the same time, making the process of charging multiple devices faster.

Battery

Turning off the Radio: Make sure the radio is off before you put the battery in or take it out. These steps can be done by turning the power/volume knob all the way counterclockwise.

How to Install a Battery

- ❖ **Aligning the Battery**: Make sure the battery is aligned to the metal frame and has a good touch with it. About one to two inches should separate the bottom of the battery from the bottom of the radio.
- ❖ **Guided Attachment:** Position the battery so that it lines up with the guide rails on the radio frame. Be careful as you slide the battery up until it clicks into place safely.
- ❖ **Securing the Battery**: Once the battery is in place, the battery latch at the radio's base locks it in place tightly, making sure it stays there.

Charging the Battery

1. Hook up the power adapter to the charger base. Plug one end of the adapter into the charger base.
2. Putting the other end of the power adapter into a wall outlet is the second step. This will power the charger.
3. Put your radio or battery in the charger's charging spot.

4. Making sure the battery's contact plates are lined up right and hitting the charger is important. Make sure the radio fits securely in the dock for charging. It means the radio is charging when the red LED light stays on.

5. While it's charging, the radio is fully charged when the charger's LED light goes green. To keep the battery from getting too full, take the radio off the charger as soon as it's charged all the way.

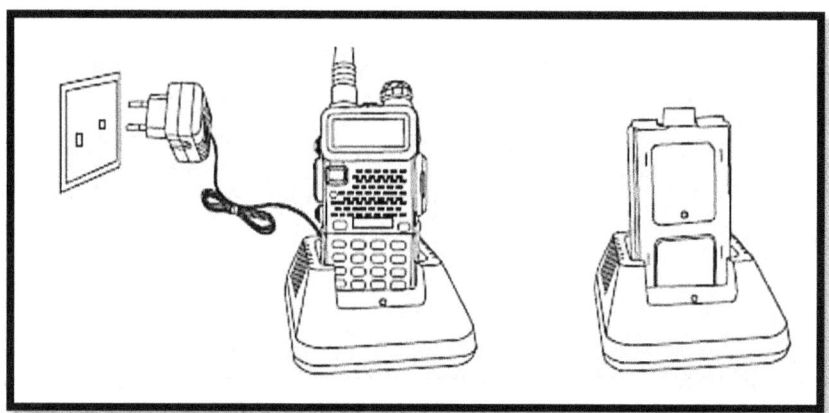

	Charger LED Codes	
Charging Status	**LED Indication**
Standby (no-load)	Red LED flashes while Green LED glows
Charging	Red LED glows solidly
Fully Charged	Green LED glows solidly
Error	Red LED flashes while Green LED glows

Taking out the battery

❖ **First, turn off the radio**. Make sure the radio is off before taking out the battery to avoid any electricity problems.

❖ **Locate the Battery Release:** The battery release should be above the battery pack. You'll need to press this to get the battery out.

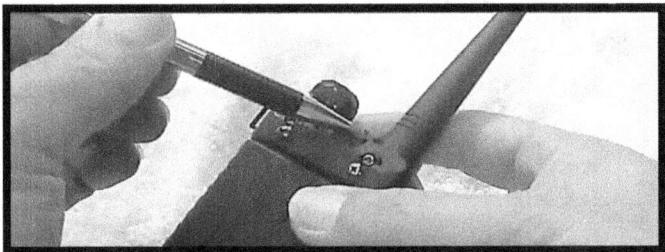

❖ **Slide Down to Release**: While moving the battery down, press the battery release button. You'll feel the battery come loose from the radio body as you slide it down a few centimeters.

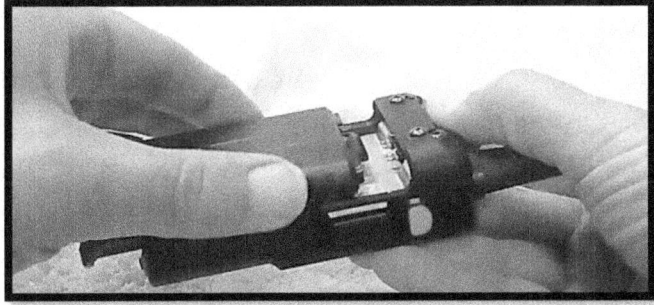

❖ **Complete Removal:** Carefully take the battery off the radio body after sliding it down a few centimeters.

Powering On and Off

To turn on a Baofeng radio, all you have to do is press and hold the orange button on the side of the device until it beeps and the screen lights up. To power off the radio, press and hold the same orange button repeatedly until you hear a beep and see the screen go blank. This simple way makes it easy to use and manage the power functions of the device.

Battery care and maintenance for your radio

Initial Charging

❖ The battery may not be charged when you get the radio for the first time. You need to charge it for four to five hours before you can use the radio.

❖ Only use batteries that have been approved by the original maker. Do not try to separate or take apart the battery.

❖ Don't put the battery near a fire or very high temperatures, and make sure to recycle used batteries the right way in your area.

How to Make Batteries Last Longer

∞ Batteries work better when they are charged in a room with normal temperatures.

∞ If you want to charge a battery that is connected to the radio faster, turn off the radio.

∞ Don't take the battery or radio out of the charger or unplug it until the charging is done.

∞ Never charge a wet battery, and if the radio's play time drops a lot, you might want to get a new battery.

∞ When working in cold places, keep a spare battery warm, like in a jacket. Dust can get into the battery contacts, so wipe them down with a clean cloth if necessary to make sure they work properly with the radio and charger.

Storage Tips

❖ Fully charge the battery before putting the radio away for a long time to keep it from getting damaged by over-discharge.

❖ Cycle the battery at least once every six months to keep it from losing a lot of power when it's not being used.

❖ Batteries should be kept in a cool, dry place that is at room temperature.

Dealing with Water Exposure

∞ If the battery gets wet, take it out of the radio, dry it with a towel, and put it in a plastic bag with dry rice that is sealed. Leave it there overnight. The rice helps the battery get rid of any extra water.

By following these tips, you can take better care of your radio's battery, making sure it works better and lasts longer while you store and use it in different situations.

Basic Controls and Functions

Adjusting Volume and Squelch

❖ Turn the volume knob on your Baofeng radio clockwise to turn it on.

❖ On the radio, find the "**Menu**" button. To get to the menu mode, press it. The menu options will be shown on the LCD screen.

❖ Click through the radio's menu options by pressing the up and down arrow keys. Find the "**Squelch**" button. Press the "Menu" button to choose it when you find it.

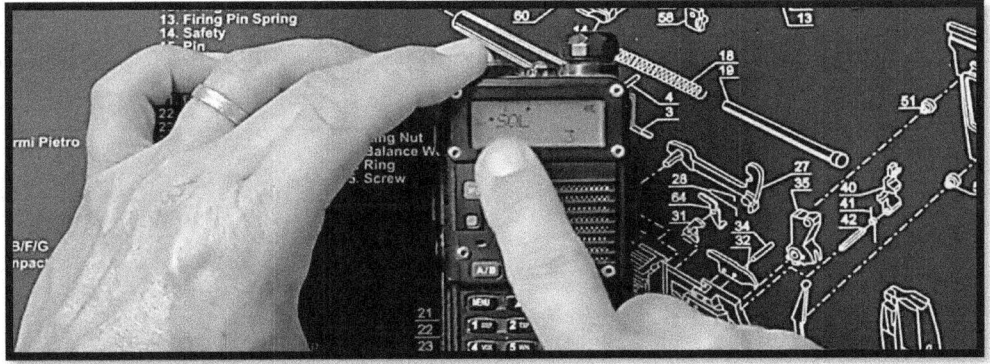

❖ Now that "**Squelch**" is chosen, you can change the level of squelch. You can change the squelch threshold value by pressing the arrow keys. Lower values will make the squelch more sensitive, which means that weaker signals will be able to be picked up but the background noise may get louder. When the value is raised, the squelch becomes less sensitive. This lowers background noise but might also block out weaker signals.

- ❖ While you change the squelch level, listen to the sound coming out of the radio's speaker or the earpiece that is connected. Find the right balances so that only important signals can be heard and the background noise is kept to a minimum.
- ❖ Press the "**Menu**" button again to save the setting when you're happy with the squelch level. The radio will stop being in menu mode.
- ❖ Listen to radio signals to test the squelch level. Make sure that you can hear the signals you want to hear and that any background noise you don't want to hear is blocked out.

Keep in mind that the best squelch level may change depending on the situation and environment you're in. Try out different settings until you find the one that makes noise reduction and signal reception work best together.

Selecting Channels and Frequencies

Baofeng radios are famous because they are cheap and can be used for many things. You can send and receive signals on a wide range of frequencies with these radios, making them useful for many situations, such as emergency contact, outdoor activities, and more.

Baofeng radios have a few steps that must be taken to choose channels and frequencies:

- ❖ First, turn on the radio and press the **VFO/MR** button to get to frequency mode.
- ❖ Use the up and down button keys to move around and pick the frequency you want.
- ❖ Press the menu button and use the arrow keys to pick either **T-CTCS or T-DCS**. This will set the tone style.
- ❖ Use the arrow keys to fine-tune and choose the tone frequency you want.
- ❖ Finally, press the menu button again to save the changes.

It's important to remember that the frequency range and way you choose a channel may be different depending on the type of your Baofeng radio.

CHAPTER 4
PROGRAMMING CHANNELS

What is meant by "programming"

On a frequency, like 146.52, people talk on the radio. A lot of different frequencies are used in radio conversations. To avoid having to remember all of them, you can set them up as channels on your radio. That is, if you set 146.52 as channel 1, all you have to do is remember to tune to that channel instead of the frequency. You don't have to type in each station every time; you can just scroll through your list.

It is called simplex when two devices talk over a single frequency. That would be single if you tune in to 146.52 and talk to someone else on that frequency. Programming is the process of adding frequencies to channels. Repeaters are automatic units that listen for broadcasts on one frequency and "repeat" them over another frequency. This is even more important when you use repeaters. There are other problems with repeaters as well. A lot of them need a certain tone from the radio, called a **CTCSS tone** or a **DCS tone** before they will play back the signal from that radio. The broadcast to that relay will not work unless the right tone is programmed into the channel. This type of communication, where repeaters use two frequencies, is called duplex instead of simplex.

Programming simplex channels into a BaoFeng

It is easier to set single frequencies by hand than repeaters, so this is a good place to begin. To begin, you should be in frequency mode instead of channel mode. In frequency mode, you can directly set frequencies. In channel mode, you can switch between channels that have already been set. Look at the screen when you turn on your BaoFeng. There are channel numbers on the right side of the screen. This means you are in channel mode. Press **VFO/MR** to go from one setting to the other.

There are two sets of frequencies on the screen: one at the top and one at the bottom. The A/B button on the BaoFeng lets you quickly switch between two frequencies. To set a frequency, you must be on the top frequency, which is shown by a small line on the left.

When you're in frequency mode and on the top frequency, use the keyboard to enter the frequency you want, such as 146.52. Like this: 146.520 To enter a frequency, you need to add zeros to the end of the number, which is what the BaoFeng does. Just type 146520; you don't need to put the decimal point. Select the frequency you want to set at the top, and then press the Menu button. You can scroll to find the **MEM-CH** menu item, but typing 27 is faster. The setting that puts channels into memory is MEM-CH. Press Menu to change that setting once you're on MEM-CH. Before you know it, the small line on the left will have moved from **MEM-CH to CH-000**, which is the usual channel.

Tip: You can leave the settings menu at any time by pressing **"Exit."** This works both before and after you save your changes.

You can put the name of the channel into the keyboard or use the arrow keys to get there. To set the frequency, find the channel you want to use and press Menu, then Exit. Press

VFO/MR to go to channel mode when you get back to the main menu. To make sure the station was set correctly, either use the arrow keys to scroll to it or type it in on the keypad.

Programming repeaters into a BaoFeng

You can add repeaters by hand once you know how to create channels in the most basic way. There are four things you need to know about repeaters: *the main frequency, the offset, the direction of the offset, and the tone*. (RepeaterBook lists tone as tone in and tone out, but for FM sources that work with the BaoFeng, they should be the same.) The frequency you listen to is the main frequency, which is the one the repeater sends on. The offset tells you what frequency the repeater is listening on, which is also the frequency you send on. Offsets are shown by numbers that are either positive or negative. This means that if the shift is -0.6 MHz and the rebroadcast frequency is 146.67 MHz, you send on 146.07. In that case, you would send on 147.27 if the difference was +0.6. **As an example, let's use W4CAT, which is a repeater around Nashville:**

- **Frequency**: 146.955
- **Offset**: -0.6 MHz
- **CTCSS**: 114.8

Do the first few steps again: make sure you're in VFO mode on the top frequency and enter the rebroadcast frequency.

Set the direction of the offset

- ❖ Press the **Menu** button
- ❖ Press **25** or move the cursor to SFT-D
- ❖ Press the **Menu** button
- ❖ Use the arrow keys to turn on, off, or +/-.
- ❖ Press the Menu button

Then set the frequency of the offset

- ❖ Press 26 or move to OFFSET if you're in the menu.
- ❖ Press the **Menu** button
- ❖ Type in the gap (type 000600 for 0.6).
- ❖ Press the **Menu** button

Lastly, the tone

❖ Press **13** or scroll to **T-CTCS** (short for send CTCSS) if you're in the menu.

❖ Press the **Menu** button

❖ Type in the tone frequency using the keyboard

❖ Press the **Menu** button

After making sure everything is right, save the frequency to a channel the same way you would for simplex. Setting the offset direction, offset frequency, and CTCSS tone should be saved to that channel. It's simple to check if everything was saved right. Find the channel for the rebroadcast. The plus sign (+-) should show up at the top of the screen to show a shift. When you press the side PTT button to send, you should see two things: CT on the left, which means a CTCSS tone is being sent; and the frequency dropping or rising to the offset. In this case, the frequency moves to 146.355 for 146.955 and a shift of -0.6 MHz.

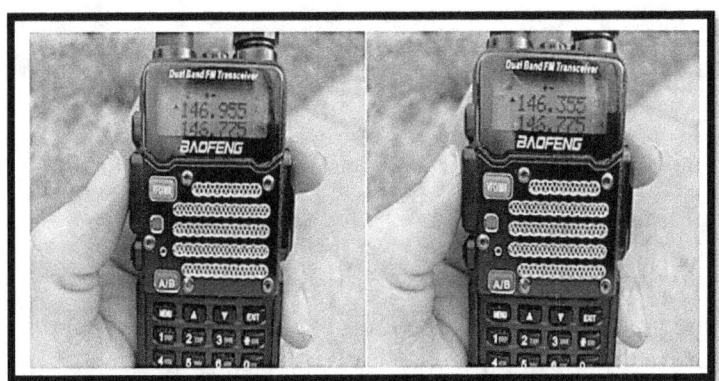

How to delete a channel

It's much easier to delete a channel than add one:

- ❖ Press the **Menu** button
- ❖ Type **28** or scroll to **DEL-CH**.
- ❖ Press the **Menu** button
- ❖ Enter or scroll to the channel you want to delete
- ❖ Press the **Menu** button

Be careful here, because there is no confirmation prompt. If you get into the menu and change your mind, press Exit before step 5.

Computer Programming

This section assumes that the Baofeng software is installed on your PC.

Attaching the Cable

Connecting the programming cable to your Baofeng radio involves a few steps to ensure a proper setup. **There are a few steps you need to take to make sure the control cable works with your Baofeng radio:**

- ❖ **Turn off the radio**. Make sure your radio is off before you connect the wire. This step stops any problems that might happen during the link.
- ❖ **Find the accessory port and open it up**. The accessory port is behind the rubber flap on the right side of the radio body. Put the cable's ends in the right place and push it down hard.
- ❖ **Connect to Computer**: Connect the cable's USB plug to your computer. Start up your machine and run the writing tools.
- ❖ **Turn on the radio**. This should be done after the wire is plugged in and the software is working.

Baofeng Software

- • **Choose a language**: The Baofeng computer tools might open in Chinese at first. Go to the second-to-right menu and choose English from the list of language options to change to English.

- **Channel Information Window**: This window usually shows up when you start up the Baofeng programming software. If it doesn't, go to Edit > Channel Information to get to it.
- **Setting up Channels**: Set up the transmission port that the cable is connected to before adding channels. After that, go to Program > Read from Radio and click "Read" to bring any channel information from the radio into the software.
- **Connection Test**: Reading information about current channels is a good way to make sure that the programming line is properly connected. The LED on the radio will flash red if the read process goes well. This means that the radio is sending data to the computer.

Channel Information Window: Column Definitions

- ❖ **Channel** -> Channel number.
- ❖ **Band** -> Displays what Frequency Band is active.
- ❖ **RX Frequency** -> Receive Frequency.
- ❖ **TX Frequency** -> Transmit Frequency. Defaults to the Receive Frequency.
- ❖ **CTCSS/DCS Dec** -> Receiver CTCSS or DCS. Defaults to OFF.
- ❖ **CTCSS/DCS Enc** -> Transmitter CTCSS or DCS. Defaults to OFF.
- ❖ **TX Power** -> Transmit power. Defaults to HIGH.
- ❖ **W/N** -> Wideband or Narrowband operation. Defaults to W for Wideband.
- ❖ **PTT-ID** -> Enables and sets the position of PTT-ID. Defaults to OFF.
- ❖ **BusyLock** -> Busy Channel Lock-out. Defaults to OFF.
- ❖ **Scan_Add** -> Add to scanner list. When enabled the channel is included in scanning mode. Defaults to ON.
- ❖ **SigCode** -> Signal Code, group ID for the channel. Defaults to 1.
- ❖ **CH-Name** -> Channel name.

CHAPTER 5

PROGRAMMING BAOFENG RADIOS WITH CHIRP SOFTWARE

Programming these radios by hand can be hard to do. CHIRP software is a good way to solve the problem. We're going to talk about how to use CHIRP to easily set up Baofeng radios so that you can communicate no matter what.

How to Block Frequencies on Baofeng Radio with CHIRP

Have you ever thought about how to keep your Baofeng radio tuned to the frequencies you want? This will give you the privacy and safety you need. These days, it's very important to be able to choose which frequencies your radio works on. You imagine being able to block unwanted signals, which would ensure that your voice and data remained crystal clear. You can read on to learn how to use the world of Baofeng radio frequency blocking and CHIRP software to take control of your radio experience.

Understanding Baofeng Radios and CHIRP

CHIRP is a piece of software that makes Baofeng radios more powerful. Acronym for "Comms Ham Radio Interoperability Programming," CHIRP is an open-source, cross-platform program designed to program different radio types. Baofeng radios are one of its main users. The great thing about CHIRP is that it can make radio broadcasting easier and better. To make it easier to send programming data between your Baofeng radio and your computer, CHIRP acts as a bridge. It gives users complete control over their radio's settings, letting them change frequencies, stations, and other important factors. This amount of control is very helpful, especially when certain frequency bands need to be stopped or given more weight. CHIRP is different because it has an easy-to-use interface that was made with both new and experienced radio users in mind. It's easy to set up Baofeng radios because it has a simple style and menu options. To use CHIRP well, you don't have to be a tech expert; it simplifies the process and saves you time and effort.

CHIRP also works with radios other than Baofeng, so it's a good choice for people who have more than one brand of radio. This interoperability makes sure that your programming skills can be used on a variety of radio types, which increases speed and makes things easier to use. Finally, Baofeng radios have made a name for themselves in the world of two-way conversation by providing a cheap and flexible option for many users. When paired with the CHIRP software, these radios are even more useful because they can be easily programmed and customized. Baofeng radios and CHIRP software make it easy and accurate to stay in touch, whether you're an amateur radio fan or a businessperson who needs to be able to communicate reliably.

Frequency Blocking: Why it Matters

Being able to block certain frequencies in the world of all the waves that move around us is comparable to playing an important part in a complex symphony where there are many different instruments. It is not simply a technological issue; it is an intelligent move that can have a significant impact, regardless of whether you are a person who enjoys listening to the radio, a business that is keeping secrets, or a government that is protecting crucial information. Let's talk about the reasons why it is vital to suppress particular frequencies and the potential consequences of leaving them open without any kind of protection.

Guarding Privacy and Security: The Why of Frequency Blocking

❖ **Privacy Preservation**: The main goal of frequency blocking is to protect privacy. Any person with a suitable device can pick up information that you send over the radio. Frequency blocking makes sure that your talks stay private and unavailable to unwanted ears, regardless of whether you're talking about private business matters or personal matters. It's like drawing the curtains over your windows to keep nosy people from seeing what you're doing inside.

❖ **Safety from spying**: In this day and age of high-tech gadgets and constant monitoring, eavesdropping is a problem for both people and businesses. By blocking certain frequencies, you make a digital wall around your contacts that keeps people who might want to listen in from getting private information. It's a digital fortress that protects the way you talk.

❖ **Integrity of the signal**: There is a chance of confusion when several radio systems are close to each other. By keeping your messages from getting mixed up with other broadcasts, frequency blocking helps keep signals strong. It is very important that your radio messages are clear and don't get messed up, whether you're a hobbyist or an expert responder.

What Are the Dangers of Open Frequencies?

❖ **Unauthorized Access**: If you leave frequencies open, people who aren't supposed to be there can get in. It's like leaving your front door wide open; anyone can come in without being asked. People listening in on your conversations without your permission can cause data leaks, privacy violations, and even security holes.

❖ **Data Vulnerability**: Data is very valuable in a world that is becoming more and more linked. If your frequencies are open, your info could be stolen or changed. Without the right frequency filtering, bad people can receive and change your data sources, hurting your company's operations and image in a way that can't be fixed.

❖ **Disturbance and Disruption**: Frequencies that are left open can be affected by disturbance from other systems or broadcasts that are not supposed to be there. This influence can mess up important messages, causing misunderstandings, slow replies, and even bad things to happen in some cases.

❖ **Compromised Security**: Secure contacts are very important to defense and security organizations. By letting people who aren't supposed to be there get into secret channels, leaving frequencies open can threaten national security and public safety. Blocking frequencies protects these important networks from threats from outside sources.

Finally, stopping certain frequencies is more than just playing with the technology; it's a strategic necessity driven by the need to protect privacy, security, and the purity of communications. Frequency blocking is the digital lock and key that protects your most valuable assets from prying eyes and possible threats in a world where information is cash. Frequency blocking is not just a choice in this linked world; it is a necessity because the risks of leaving frequencies open are too big.

Getting Started with CHIRP: Your Path to Radio Programming

Precision and adaptation are important in the world of two-way radio contact, which is always changing. Starting the process of setting up CHIRP on your computer is like getting your Baofeng radio to work at its best. Here is where the magic starts, where frequencies match your wants, and where easy contact is the main event. Let us go over the steps needed to get started with CHIRP, making sure to stress how important it is to have a suitable programming line.

Step-by-Step Guide to CHIRP Setup

❖ **Downloading and setting up**: Start your CHIRP trip by going to the official CHIRP website *(https://chirp.danplanet.com/projects/chirp/wiki/Home)* and getting the most up-to-date software that works with your computer, whether it's **Windows, macOS, or Linux**. After the download is done, follow the on-screen steps to run CHIRP. Anyone can easily get started with this process, even if they don't know much about computers.

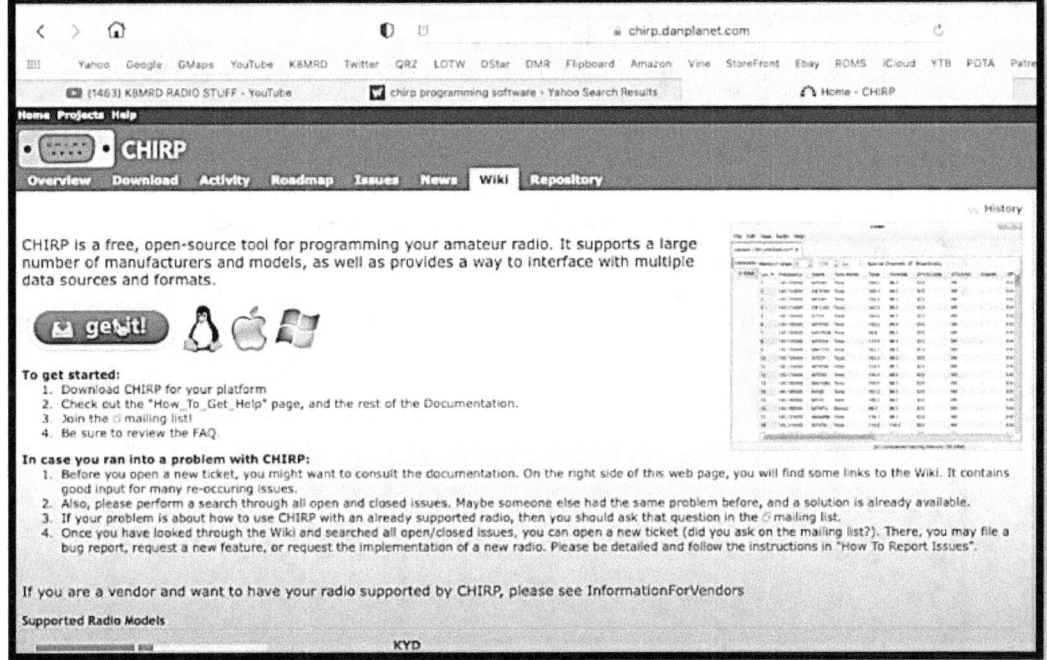

❖ **Connecting Your Radio**: Now that CHIRP is set up, it's time to make the important connection between your computer and your Baofeng radio. It is important to use a suitable programming cable because it is how the data gets from one device to another. Connect one end of the cord to the USB port on your computer and the other end to the programming port on your Baofeng radio. Make sure that both devices are turned on and ready to talk.

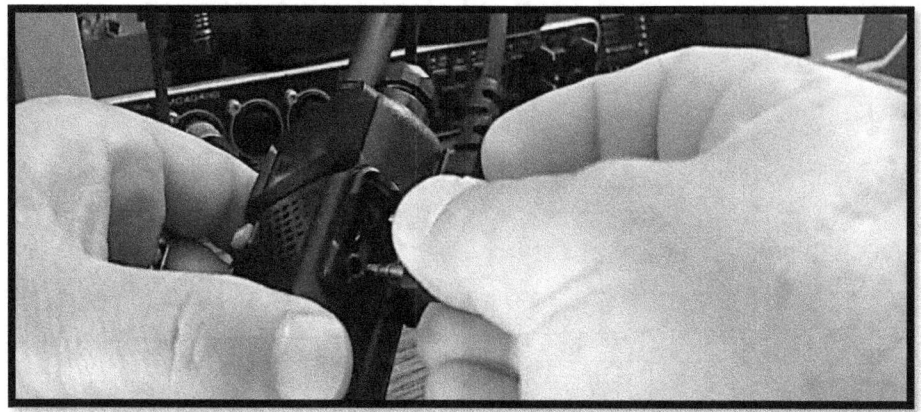

- ❖ **Radio Identification**: When you open CHIRP, you'll see a clean, easy-to-use layout. The next step is to figure out what kind of radio you have. Click **Radio** at the top bar and select "**Download from Radio**".
- ❖ Pick out the brand from the "**Radio**" menu; in this case, it's "**Baofeng**." Pick out the Baofeng radio type you want from the list given. This step is very important because it makes sure that CHIRP fits the needs of your radio.

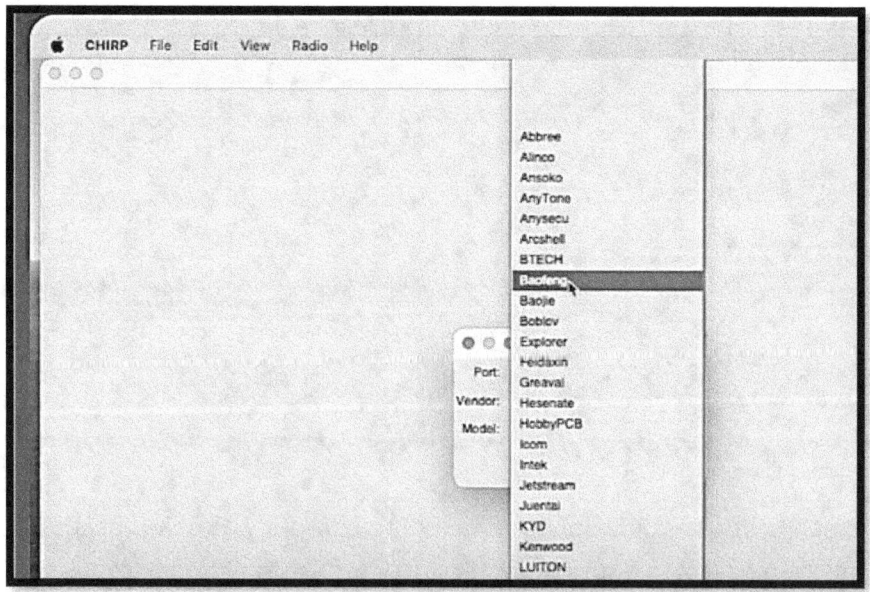

- ❖ **Read and Save Existing Configuration**: Use the "**Read from Radio**" option in CHIRP if you've already set your Baofeng radio or want to keep it in the configuration it's in now. This takes your radio's current settings and shows them on the software screen. Before making any changes, make sure to save this setting as a backup.

❖ **Setting the Frequencies**: This is the most important part of the process. You can easily set the frequencies you want with CHIRP's easy-to-use interface. Make new channels, give them names that explain what they do, and enter the frequency numbers. You can also change things like the broadcast power, the delay, and the tone settings. CHIRP gives you a picture of your channel lineup, which makes it simple to arrange and change.

❖ **Uploading to Radio**: Once you've carefully set up your frequencies and channels, it's time to put this information on your Baofeng radio. After using CHIRP, all you have to do is click on "**Upload to Radio**" and watch as your radio takes on your interests. After this step, your Baofeng radio will be ready to use for conversation.

The Crucial Role of a Compatible Programming Cable

When it comes to this complicated dance of sending data and customizing the radio, the programming wire is the star. Any old wire will not do; this is the one that connects your radio to CHIRP on your computer. **Here's why you need a setting wire that works with your device:**

❖ **Smooth Connections**: A programming cable that works with your computer and Baofeng radio will link smoothly. It lowers the chance of contact problems or data transfer mistakes, which makes the computing process, go smoothly and quickly.

❖ **Accuracy and Precision**: Accuracy is important when working with radio frequencies. A suitable connection is designed to send data accurately, so you can be sure that your radio's settings are exactly what you want them to be.

❖ **Data Integrity**: Your data must be correct. A good programming line keeps data from getting damaged while it's being sent. This is especially important when working with important frequencies or channels for emergency contact.

❖ **Saves time and effort**: Trying to get by with a connection that doesn't work can be annoying and take a lot of time.

❖

❖ A suitable cord makes the whole process of setting easier and f aster, saving you time and effort.

Connecting Baofeng Radio to CHIRP

A key step toward personalized radio contact is getting your Baofeng radio to work with the powerful CHIRP software. When you enter the world of connection, you start a trip that combines hardware and software, letting you shape your radio's features to suit your needs. This part will show you how to connect your Baofeng radio to CHIRP so that the process goes smoothly and quickly. It will also cover common connection problems and give you advice on how to fix them.

Walking through the Connection Process

❖ **Start CHIRP on your computer**. Pick up the "**Radio**" menu and pick "**Download from Radio**." CHIRP will then ask you to choose your radio's maker, which is "Baofeng," and then pick your Baofeng radio type from the list that is given. This step makes sure that CHIRP is set up to work with your radio.

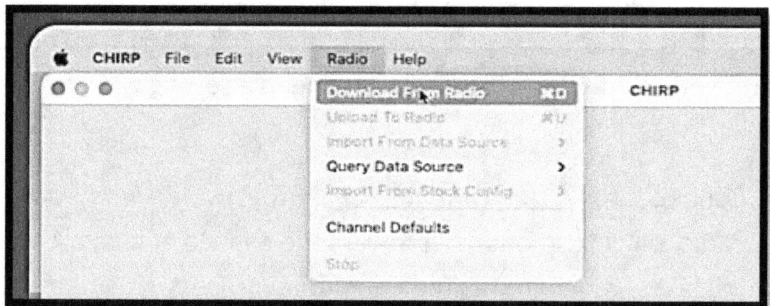

❖ **Communication Protocol**: CHIRP will then try to connect to your Baofeng radio and start talking. In CHIRP's settings, make sure that the right COM port is chosen. If you're not sure which COM port to use, look in the device manager on your computer for the current port that is linked to your radio.

❖ **Read from the radio**: Click "**OK**" to move on after the link is made. Now, CHIRP will read your radio's current settings and show them on the software screen. This step is very important if you want to save a copy of your current settings or make changes.

❖ **Change and Upload**: Once you see your radio's setup in CHIRP, you can change and tweak the settings to your liking. You can change frequencies, add or remove channels, and set other settings as needed. When you're happy with your changes, click "Upload to Radio" to send the new settings to your Baofeng radio.

Troubleshooting Tips for Common Connection Issues

Connecting your Baofeng radio to CHIRP is usually a simple process, but sometimes things can go wrong. For common link problems, here are some ways to fix them:

- ❖ **Choose the Right COM Port**: Check that you've chosen the right COM port in CHIRP's settings. If the COM port doesn't match, contact can't happen.
- ❖ **Cable Integrity**: Make sure that the quality of your television line is good. Cables that are broken or malfunctioning can sometimes make it hard to communicate. If you think this is the problem, try a different connection.
- ❖ **Installing Drivers**: Make sure that your computer has the drivers that your programming cord needs. Driver software is often provided by manufacturers and needs to be loaded for transmission to work.
- ❖ **Radio Compatibility**: Make sure that CHIRP works with your type of Baofeng radio. This can lead to problems if the software doesn't recognize your radio type.
- ❖ **Radio Power**: Make sure your Baofeng radio is turned on while you're connecting it. The radio and CHIRP can't talk to each other without power.
- ❖ **USB Port**: Connect the cord to a different USB port on your computer to see if that works. Problems with USB ports can sometimes make it hard to communicate.

Backing up Radio Configuration

Backing up Radio Configuration: Safeguarding Your Radio's Digital DNA

When you talk on the radio, you carefully choose each frequency, channel, and setting to meet your specific needs. In the same way, your radio's personality is shown by its digital fabric or setup. Your tastes, your communication needs, and your working skills all come together in this digital DNA. It's a gem that should never be in danger of being forgotten. his is where the importance of backing up your radio's setup using CHIRP comes into play, and in this part, we'll talk about why it's so important and how to do it correctly.

The Importance of Creating a Backup

❖ **Preserving Customization**: Your Baofeng radio is a flexible tool, and the way it's set up shows how you use it. This customization shows that you know what you need, whether you've carefully set up frequencies for emergency calls or channels for fun activities. Making a backup protects this valuable digital board from losing any data that comes out of the blue.

❖ **Security Blanket for Data Loss**: Technology isn't always reliable, and data loss can happen for many reasons, such as software bugs or events that were not planned for. A copy of your radio's settings is like a safety blanket; it gives you peace of mind that you can get your radio back to its best state if you lose data.

❖ **Efficiency in Replication**: It's easier to share your setup with other people or own multiple Baofeng radios if you have a backup. You can quickly copy settings across devices instead of starting from scratch. This saves time and makes sure that everything is the same.

A Guide to Saving Your Current Configuration with CHIRP

❖ **Start CHIRP**: To begin, start the CHIRP software on your computer and make sure that the appropriate programming wire is plugged into your Baofeng radio.

❖ **Read from Radio**: In CHIRP, go to the "**Radio**" menu and choose "**Download From Radio.**" This will allow CHIRP to connect to your Baofeng radio and get its current settings. This is the setup that you want to keep safe.

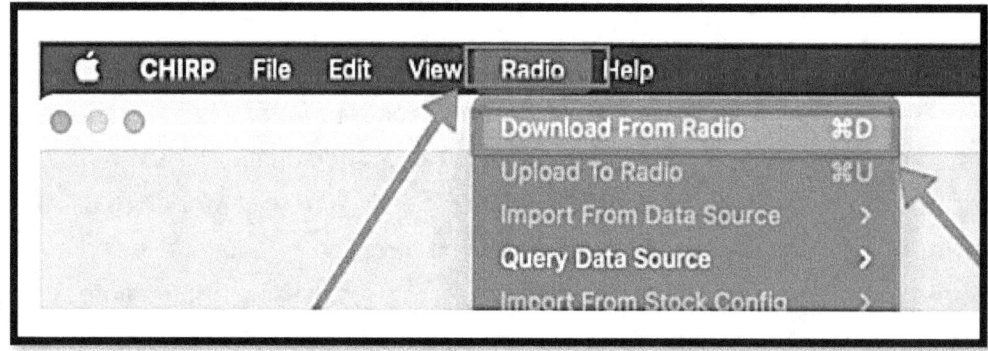

❖ **Saving the Configuration**: As soon as the CHIRP screen shows you your radio's setup, go to the "**File**" menu and choose "**Save As.**" Pick a place on your computer to save the backup file and give it a name that makes sense. Make sure the file type is set to ".**img**" so that it can be restored in the future.

❖ **Finish the Process**: Once you've chosen the place and name of the backup file, click "Save" to make it. These are the files that CHIRP will use to save all of your radio's settings, stations, and frequencies.

❖ **Verification**: To be sure the backup worked, you can go to where it was saved on your computer and make sure the backup file is there. You should check the security of your saved files regularly to be sure they are still valid.

CHAPTER 6
IDENTIFYING FREQUENCIES TO BLOCK

Identifying Frequencies to Block: Navigating the Spectrum

When it comes to radio transmissions, being able to carefully block frequencies is one of the most important skills for controlling the noise of electromagnetic waves around us. Finding the frequencies to block is one of the most important things you can do to improve your radio scene, whether you're a radio fan trying to keep the signal strong, a business trying to keep private data safe, or a government agency making sure everyone follows the rules. Here, we'll talk about how to study and find frequencies that need to be blocked; all while keeping a close eye on the legal issues that come up in this complicated area.

Researching and Identifying Frequencies to Block

❖ **Frequency Databases**: Detailed frequency databases are one of the best ways to find frequencies to block. These databases, which are usually kept up to date by government bodies or industry groups, hold a lot of useful details about licensed frequencies, assignments, and license users. In the US, the **Federal Communications Commission's (FCC) Frequency Database** is a useful website. In New Zealand, the **Radio Spectrum Management** is a useful website.

❖ **Scanner and Spectrum Analyzer Tools**: Buying scanner and spectrum analyzer tools can help you find frequencies that are actively being used in your area. You can scan the radio spectrum with these gadgets to find signals and frequencies that might need to be blocked. They give real-time information on signal strength, frequency, and modulation, which helps with the process of identification.

❖ **Collaboration with Regulatory Authorities**: In some cases, working together with legal officials can be helpful, depending on your needs and the rules in your area. These groups can advise on things like frequency assignments, license requirements, and possible interference problems. On top of that, they can help you find frequencies that could be dangerous or illegal.

❖ **Frequency Monitoring Software**: Using specialized frequency monitoring software can make the process of identification go more quickly. You can keep an eye on and record radio frequencies in your area with these software programs. This will help you find patterns and trends. They are especially useful for groups that want to make sure they follow radio laws.

Legal Considerations When Blocking Frequencies

❖ **Licensing and Authorization**: It's important to check if you are legally allowed to stop any frequency before you do so. A lot of radio frequencies are licensed and assigned by government agencies. Interfering with legal broadcasts can get you in trouble with the law. Always make sure that you have the right permissions or licenses to block certain frequencies.

❖ **Safety and Emergency Services**: Some frequencies are set aside for contact between emergency services and the people. Putting these frequencies off can put people's safety at risk and have bad results. Protecting these important routes must be a top priority, and rules must be followed to keep them safe.

❖ **Interference Mitigation**: Even though blocking frequencies to avoid interference is a reasonable purpose, it is necessary to do so within the confines of the law. Some countries may require you to coordinate your efforts to mitigate interference with regulatory authorities to prevent unintended breaches.

❖ **Regarding privacy and security**, it is a good idea to block frequencies to keep conversations private and safe. But it's very important to know about any privacy laws or rules that might apply. Legal problems can happen if someone listens in on conversations or blocks frequencies without permission.

Programming Blocked Frequencies

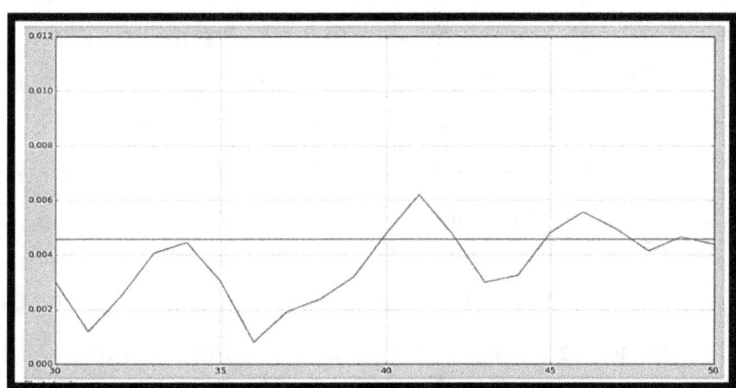

Programming and managing blocked frequencies is like using a conductor's baton to put together a symphony in the complicated world of radio communication. It's a very careful process that needs accuracy, attention to detail, and knowledge of how hardware and software work together in complex ways. In this part, we'll talk about how to use the flexible CHIRP software to set blocked frequencies. We'll stress how important accuracy is in this process.

Detailed Instructions on Adding Frequencies to CHIRP for Blocking

❖ **Start CHIRP**: To begin, start the CHIRP app on your computer. Make sure that the programming wire that came with your Baofeng radio is compatible with this one. This is how you'll send your set frequencies.

❖ **Download from Radio**: To download from a radio, go to the "**Radio**" menu and choose "**Download from Radio**." This will get your radio's current settings and let you see and change them.

❖ **Choose Frequencies to Block**: Carefully go over the frequencies you want to block. These could be frequencies that are interfering with other signals or sending unwanted signals. They could also be frequencies that need to be limited for practical reasons. Write down these frequencies so you can find them again.

❖ **Input Frequencies**: In CHIRP, find the channel where you want to add a filtered frequency and click on it. If the cell is called "**Frequency" or "Rx Frequency**," click on it and enter the frequency you want to stop. Make sure you enter the frequency correctly, in megahertz (MHz), with the decimal point in the right place.

❖ **Adjust Channel Settings**: Depending on your needs, you may need to adjust other channel settings, such as tone delay, broadcast frequencies, and channel names. Make sure you carefully set these parameters so they work for you.

❖ **Repeat for Each Frequency**: Create a separate channel for each blocked frequency if you have multiple frequencies to block. With this level of detail, you can precisely control and keep an eye on blocked frequencies.

❖ **Upload to Radio**: Once you've added and set up all the frequencies you want to block, it's time to send your Baofeng radio the new settings. To set your radio to block certain frequencies, go to the "Radio" menu and choose "**Upload to Radio**." CHIRP will then send the new setup to your radio.

Testing and Verifying Frequency Blocking

To keep your frequencies safe and in control is like keeping your castle safe. Precision is very important, and any weak spot could let your communication network down. As we start to try and confirm the frequency blocking on your Baofeng radio, we embark on a journey that demands not only technical finesse but also an unwavering commitment to airtight security.

Explaining the Process of Testing for Frequency Blocking

❖ After using the **CHIRP** software to set the frequencies you want to block into your Baofeng radio, the first thing you need to do to test it is to start a frequency shutdown. To do this, you need to set up your radio so that it doesn't let you use the blocked frequencies.

❖ Do a transmission test to make sure that the blocked frequencies can no longer be reached. Try sending or receiving messages on the frequencies that you have set to be blocked. You should either hear a solid silence or an error message that says "*access is denied*" if your frequency blocking is working right.

❖ Use tools or features built into your Baofeng radio for signal tracking to check for unwanted or blocked frequencies. Keep an eye on the radio spectrum to see if there are any illegal messages or broadcasts on the frequencies that have been blocked. If there is action on these frequencies, it should be looked into further.

❖ Use a scanner or spectrum analyzer to look across the radio spectrum and make sure that the frequencies you set to be blocked are still not being used. These gadgets give you real-time information about signal behavior, which lets you, finds any problems.

Troubleshooting Issues during Testing

❖ **Mistaken Positives**: While testing, you might come across fake positives, which are times when real signals are wrongly thought to be blocked frequencies. To lessen this effect, make sure that your programming settings are correct. This includes making sure that tone squelch and broadcast frequencies are set properly. Change these settings as needed to get rid of fake results.

❖ **Interference:** Interference from nearby frequencies or outside sources can sometimes look like action on a frequency that is blocked. To fix this, you might want to change how sensitive your radio's antenna is or add more filters to cut down on crosstalk.

❖ **Updates for the software**: Sometimes, out-of-date firmware can cause strange behavior during frequency blocking. Make sure that the software on your Baofeng radio is up to date. Manufacturers often release changes that fix problems and make the equipment work better.

❖ **Read the documents**: If you keep having problems while trying, read the documents that came with your Baofeng radio and the CHIRP software. Manufacturers often offer repair tools and Frequently Asked Questions (FAQs) that can help you figure out how to fix technology issues.

Creating and Saving New Configuration

Creating and Saving New Configuration: Safeguarding Your Communication

When it comes to radio contact, being able to make and save a new setup for your radio is like drawing up a plan for safe and effective talks. Precision and attention to detail are needed to make sure that your frequencies, channels, and settings are perfectly tuned to your needs. This part will start a trip that will show you how to make and save your newly designed arrangement. Along the way, we'll talk about why it's important to keep a record of your radio's setup.

Guide to Saving the Newly Programmed Configuration

❖ **Connect Your Radio**: To start, use a suitable programming wire to connect your Baofeng radio to your computer. Make sure that your radio is turned on and ready to be programmed.

❖ **Launch CHIRP**: Start up the CHIRP app on your computer. There is a link between your computer and your radio called CHIRP that lets you set up and control settings easily.

❖ **Download from Radio**: In CHIRP, go to the **"Radio"** menu and choose **"Download from Radio."** This will get the current setup from your radio and show all of the settings, stations, and frequencies on your computer screen.

❖ **Change the setup:** Change the setup as needed by adding or deleting frequencies, changing the silence settings, or fine-tuning the amounts of broadcast power. This step makes sure that your radio is perfectly suited to your conversation needs.

❖ **Save Configuration**: When you're done modifying your configuration to your liking, click the "File" menu and select "Save As." Give your configuration file a name that is easy to remember and that describes the specific situation or reason for which it was created.

Why keeping track of the configuration is a good idea

❖ **Instant Restoration**: If you lose data or change the settings by mistake, having a saved setup file lets you quickly get your radio back to how it should be. This cuts down on downtime and keeps contact going at all times.

❖ **Scenario-Based Profiles**: You can make scenario-specific profiles by keeping track of multiple saved setups. Having these profiles on hand makes it easier to switch between radio networks or meet different contact needs.

❖ **Documentation and Compliance**: Keeping a record of your setups is a good way to show that you're following operating and legal rules for legal or business reasons. It shows that the rules have been followed and can be very important in some fields.

❖ **Efficient and Consistent**: Saved settings help make operations more efficient and consistent. They keep you from having to change the settings on your radio by hand every time you use it, so you can focus on talking instead of setting it up.

❖ **Accidental Damage**: If something bad happens, a saved setup can be used as a safety net. It makes sure that your radio is ready to go quickly in emergencies where time is of the essence.

❖ **Simple Sharing**: If you work with others on a team or in a group, sharing saved settings makes it easier to make sure that everyone is on the same page. It makes things more consistent and lines up how people talk to each other.

Additional Tips and Best Practices for Optimal Baofeng Radio Performance

While you are learning how to use Baofeng radios and CHIRP programming, there are some other tips and best practices that can make your radio communication even better. Not only do these insights improve speed, but they also make sure that your frequency block choices stay strong and stable.

Tips for Optimizing Baofeng Radio Performance

❖ **Antenna Matters**: The antenna is what connects your radio to the outside world. You might want to buy high-quality antennas that are made for the frequency bands you want to use. A good antenna can make both receiving and sending much better.

❖ **Battery Management**: It's very important to be able to handle batteries well, especially on long tasks or trips. To avoid being without power when you least expect it, bring extra batteries with you or buy a high-capacity battery pack.

❖ **Getting regular firmware updates**: Baofeng will send you software updates to keep your device up to date. Most of the time, these changes fix bugs, make the radio run better, and add new features that can make it more useful.

❖ **Channel Organization**: Sort your channels correctly by giving them names that are clear and explain what they do. This makes it easier to change frequencies, especially when things are moving quickly.

❖ **Think about getting an extra speaker-microphone**. This can help with sound quality and make things easier, especially in noisy places. Also, you can keep your radio safe while talking on the phone.

Maintaining Frequency Block Settings

❖ **Regular Audits**: Check and double-check your frequency block settings every so often. Make sure that they still meet the law and marketing needs of your business. This is very important because rules and frequency assignments can change over time.

❖ **Documentation**: Keep thorough records of the frequencies you've blocked. Include the reason for the block, the date it will happen, and any important legal references. For safety and fixing, this paperwork can be very important.

❖ **Check and Test**: Make sure that your blocked frequencies are blocked regularly. This proactive method keeps your radio safe from broadcasts you don't want.

❖ **Update as Needed**: If your contact needs to change or if you start getting interference on frequencies that weren't blocked before, be ready to change your block settings.

❖ **Backup setups**: You should make backups of your frequency block sets just like you do with your radio setups. This gives you a backup in case you make changes by mistake or lose info.

By using these extra tips and best practices when you use your Baofeng radio, you can get the most out of it while still protecting your frequency block settings. Remember that radio transmission is both a skill and a duty. To be successful in this ever-changing area, you need to stay informed and ready. These tips can help you find your way around the airwaves with confidence and accuracy, no matter how long you've been a radio fan or how new you are to it.

How to program baofeng UV-82

To program your Baofeng UV-82 on MAC you will need two things:

- A Baofeng programming cable
- Chirp programming software

Chirp is open-source ham radio programming software. It has a free version that is accessible to all hams or anyone who needs to program a Baofeng Radio. You will need it to set up your Baofeng UV-82, so please get it from the Chirp website. Chirp can be used with just a double click, so you don't have to install it. As Chirp is not signed, your Mac OS may decide to stop it. If this happens to you, all you have to do is allow it in the system settings.

Setting up your Baofeng UV-82 on a Mac is a very simple process. This is what needs to be done:

- Read the radio's configuration
- Modify or add your parameters
- Write the modified configuration to the radio

You will be able to set your radio if these steps go smoothly. Let us break it down.

How to program baofeng UV-82 on Mac – read the radio

It is always a good idea to read the radio's setup before programming it. Most tools for writing radios won't even let you write something to them without first reading it. For Chirp to be different from the others, we must first read the radio before we can change it.

You will need to do the following to read the radio's mode:

- Go to **Radio > Download from Radio** after starting Chirp.

- Pick out the serial device that comes with the setting wire, and then choose Baofeng as the seller and UV-82 as the radio model.
- Click ok to read the radio's configuration.

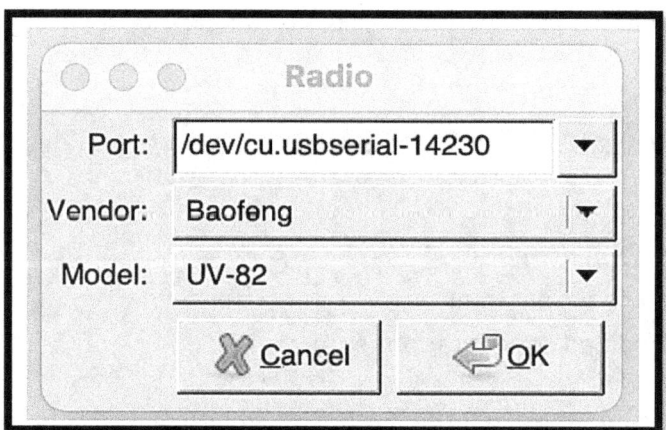

- Do what it says on the screen and click "**OK**."

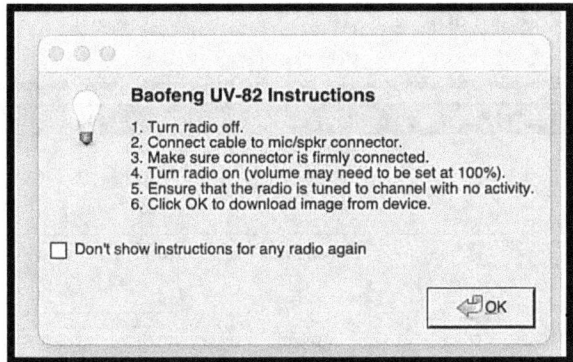

- If everything is right, you will see this window popping up.

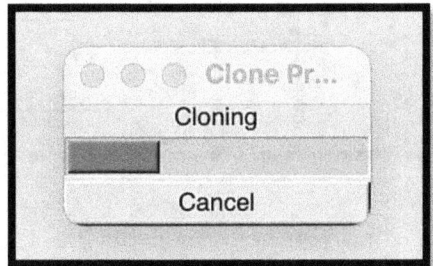

- **Go to Memories**: You can then add all the channels you need in Memories.

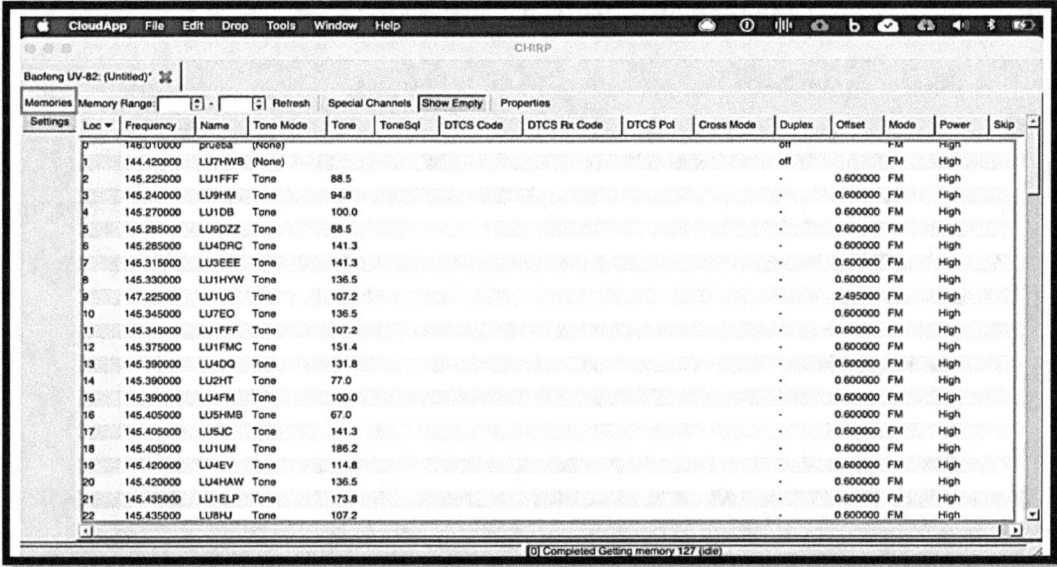

- After that, change the radio's settings to suit your wants.

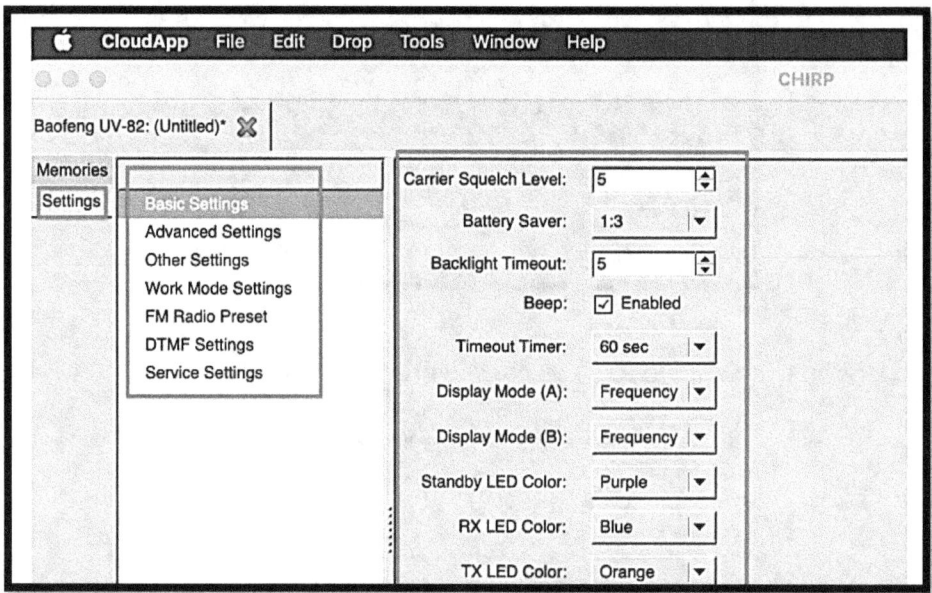

- After everything is finished, we need to send the settings to the radio. Click on Radio in the menu and then click on **Upload to Radio**.
- Type, make, and model all the same. To go on, click "**OK**." Do what it says, and then click "**OK**" to move on. Wait until the copying is done. That's it. You can now use CHIRP to set your Baofeng UV-82 from a MAC OS.

Note: It is still the same steps and procedures with Windows.

By programming Baofeng radios with CHIRP software, you can get the most out of them and communicate clearly in a wide range of situations. Anyone who likes the outdoors, works as an emergency responder, or just likes radios can program Baofeng radios with CHIRP to make them more useful and give them more customization options.

Radio to Radio Cloning

The Baofeng UV-82 has an amazing feature called **"radio cloning**," which lets settings be copied seamlessly between devices. This feature makes it easy to move settings from one radio to another, making sure that all of them are set up the same. A certain connection is used to connect a Reference (Master) radio to a Copy (Slave) radio, and then the data is copied from one device to the other.

Here's a step-by-step guide for cloning radios:

❖ Connect the adapters for the copying cable to the auxiliary cable ports on both the Reference and Copy radios to use it.
❖ Turn on the Copy radio, which is what the copied settings are meant to go to.
❖ Hold down the [**MONI**] button and turn on the Reference radio. This is the source from which the settings will be copied.
❖ On the Reference radio, the screen should say "**COPYING**," and if the link works, the LED will flash red to show that data is being sent. At the same time, the Copy radio's LED will flash green to show that data has been received.
❖ When the LEDs on both radios go out, they will restart, finishing the copying process. The action was successful.

Automatic Number Identification (ANI)

In dispatch settings, an important method makes it possible for radios to instantly identify themselves to the dispatcher. Automatic Number Identification (AIN) or PTT-ID is the name of this system, which requires radios to send a data burst with their ID code at the beginning or end of a communication. Additionally, the Baofeng UV-82 uses DTMF signal to make ANI possible.

To set up the ANI/PTT-ID Code, follow these steps:

1. To use the Baofeng PC program, connect the radio to a computer and start it up.
2. To open the **DTMF Encode/Decode** box, go to the **Edit menu** and choose **DTMF**.
3. To open the **Read from Radio** window, go to the **Program menu** and select Read from Radio.
4. Press the **Read** button. The radio's status LED will flash red to show that data is being sent.
5. Find the **ANI Code** box and type in the correct ANI code information.
   • In the **DTMF Encode/Decode** window, you can enter up to 15 group ID numbers from the list on the left. In the station Information field, these can be given to each station separately.

6. Check the box next to "**Press PTT to Send**" to send the ID before a normal message.

7. When you check the "**Release PTT to Send**" box, the ID will be sent after a normal message.

8. Go back to the Program menu and choose Write to Radio. This will bring up the Write Data to Radio window.

9. Press the "**Write**" button. The radio's status LED will flash green to show that it is receiving data.

If the radio is still linked to the PC and the software is still working, do these extra steps to fully enable the ANI settings:

Procedure: Enabling/Disabling/Configuring ANI Settings

❖ To open the **Optional Features** box, go to the **Edit** menu and choose **Optional Features**.

❖ To open the **Read from Radio** window, go to the Program menu and select Read from Radio.

❖ Press the Read button. The radio's status LED will flash red to show that data is being sent.

❖ The Optional Features window has a drop-down list for **PTT-ID**. This lets you choose where the ANI data burst should happen: **BOT** (Beginning of Transmission), **EOT** (End of Transmission), or **BOTH**. Choose **OFF** from the drop-down menu to turn off ANI completely.

❖ Once more, go to the **Program menu** and choose **Write to Radio**. This will bring up the **Write Data to Radio** window.

❖ Press the "**Write**" button. The radio's status LED will flash green to show that it is receiving data.

If you follow these steps, the radio should be properly set up for ANI functions.

Saving and Naming Channels

❖ On the radio, press the **Menu** button.

❖ Use the up and down button keys to get to the Memory option, and then press the Menu key.

❖ Use the arrow keys to choose which memory channel you want to save, and then press the Menu button.

❖ Pick the channel's frequency that you want and press the **Menu** button.

- ❖ Use the arrow keys and the Menu key to get to the Name field.
- ❖ Use the keyboard to give the channel the name you want, then press the Menu button.
- ❖ Press the Exit button to save the channel.

You can only show one of the Name, Frequency, or Number of a memory spot at a time on Baofeng UV-5R radios with a two-line display.

This is true for each of the two memory slots. To work around this limitation:

- ❖ Program the A and B settings separately.
- ❖ Set channel A to display the name and channel B to display the frequency
- ❖ Configure the receiver to listen only to the monitored channel to avoid potential issues.
- ❖ Use your preset station (like station 12) on both A and B at the same time in channel mode. In this case, A will show the nickname for your channel name and B will show the frequency. This way is a fix that lets you see both the channel name and the frequency at the same time, giving you the information that you need even though the display is limited.

Power-On Message

The Baofeng PC program is the only way to change the power-on message on your Baofeng radio.

This can be done by following these steps, assuming that the Baofeng software is already set up and working and that your Baofeng radio is linked to your PC:

- ❖ Make sure your Baofeng radio is connected to your computer before you open the Baofeng PC program.
- ❖ On the software's menu bar, click the "**Other**" button. When you do this, a text box called "**Other**" will show up.
- ❖ In the chat box's "**Power on Message**" area, there are two text fields that show lines that will show up on the radio's LCD screen. Type the words you want to use in these areas.

❖ After typing the text, press the "Write" button on the software. This move saves the changes you made to the radio and makes them work.
❖ On the real Baofeng radio, make sure that **"MSG"** is selected in menu item 38.

It is important to know that the Baofeng UV-82 radio can only show 7 characters on each line on its LCD screen. Make sure that the text you typed fits within this range.

CHAPTER 7
RADIO ETIQUETTE AND COMMUNICATION PROTOCOLS
Understanding Radio Language and Terminology

Radio language and terms are very important for communicating clearly, especially when using Baofeng radios, which are popular with amateur radio operators, first responders, and people who like being outside. Baofeng radios are known for being affordable and flexible, but to get the most out of them, you need to know the specific terms and words used for radio conversation. Let's look at some important radio slang terms that have to do with Baofeng radios.

- ❖ **Frequencies and Channels**: It is very important to understand frequencies and channels. Frequencies are the radio waves that are used for transmission, and channels are groups of frequencies that have already been decided upon. Users of Baofeng radios can set certain frequencies as channels so that they are easy to find.
- ❖ **CTCSS and DCS**: Continuous Tone-Coded Squelch System (CTCSS) and Digital-Coded Squelch (DCS) are two features that many Baofeng radios have. These are digital codes or tones that are not noticeable but help block out unwanted signals so that only radios set up with the same CTCSS or DCS can hear each other.
- ❖ **Simplex and Duplex:** When it comes to radio contact, simplex means that transmission and reception happen on the same frequency, while duplex means that transmission and reception happen on different frequencies. Baofeng radios can work in both solo and duplex a mode, which gives you options for a variety of situations.
- ❖ **Push-to-Talk (PTT):** The PTT button is an important part of radio transmission. Like many other radios, Baofeng radios have a PTT button that must be pressed when sending and let go of when hearing. To keep talks from crossing over, it's important to know how to use the PTT properly.
- ❖ **Squelch**: This is a circuit in radio that turns off the sound when there is no information coming in. Users of Baofeng radios can change the silence levels to get rid of background noise while still being able to hear important messages.

- ❖ **Repeater Operation**: Baofeng radios can be set up to work with repeaters, which are gadgets that pick-up messages on one frequency and send them back on a different frequency at the same time. This greatly increases the number of people who can communicate. It is important to understand the repeater distance and tone settings to use the repeater correctly.
- ❖ **Emergency Codes and Procedures**: Standardized codes and procedures are very important in case of an emergency. Baofeng radios have emergency channels and tools that let users quickly get to set frequencies for contact in an emergency.
- ❖ **Radio Protocol**: It is important to follow set radio procedures so that transmission works well. This means sending clear messages and to the point, spelling out words using phonetic alphabets, and staying away from useless talk on shared frequencies.
- ❖ **Battery Care**: Most Baofeng radios come with batteries that can be charged again and again. Knowing how to handle batteries, such as how to charge them, how long they last, and what backup power options are available, is important for staying connected in the field.
- ❖ **Licensing**: Depending on where you live, you may need a legal license to operate a Baofeng radio or any other radio broadcast. Understanding and following the rules for getting a license is not only the law, but it also makes sure that radio use is done safely and without interruption.

How to Use CTCSS and DCS Tones Correctly

Continuously Coded Squelch System (CTCSS) and Digital-Coded Squelch (DCS) are two types of two-way radio communication that protect privacy and make it less likely that someone will mess with the conversation without meaning to. These functions are often found on Baofeng radios because they are popular with amateur radio users. Mastering the correct use of CTCSS and DCS tones can greatly improve the efficiency of conversation.

1. A Quick Look at CTCSS and DCS:
- • **CTCSS (Continuous Tone-Coded Squelch System):** To send and receive messages, CTCSS uses tones that can't be heard. A different tone is given to each user or group of users, and the radio will only let sound through when it gets a signal with the right CTCSS tone.

- **DCS (Digital-Coded Squelch):** In DCS, digital codes are used instead of analog tones, making it a more modern method. Like CTCSS, each person or group has their code, and the radio will only open the silence when it receives the right digital code.

2. How to set CTCSS/DCS on Baofeng radios:
- Press the **MENU** button on the radio to get to the menu.
- To get to the "**CTCSS**" or "**DCS**" setting, press the arrow keys.

- Choose the **CTCSS** tone or DCS code that goes with the group or channel you want to talk to.
- Save the changes so that they take effect.

3. Picking the Right Tones and Codes:
- Work together with your group: To make sure everyone can talk to each other; make sure everyone in your contact group is using the same CTCSS tone or DCS code.
- Look at frequency charts. There are many places where you can find frequency charts with the CTCSS tones and DCS codes that go with them. This guide helps you pick tones or codes that aren't used very often in your area.

4. Why using CTCSS and DCS is a good idea:
- **Less disturbance**: CTCSS and DCS help get rid of signals that don't have the right tone or code, which makes it less likely that other users on the same channel will cause disturbance.
- **Privacy**: You can make a private line of conversation on a shared frequency by using different tones or codes.

5. Common Pitfalls and Best Practices:
   - **Compatibility**: Make sure that all of the radios in your group can handle CTCSS and/or DCS and that user can set the right tones and codes.
   - Before depending on CTCSS or DCS in an emergency, you should test your radios to make sure they are set up correctly and can interact well.

6. Emergency Communication:
   - In an emergency, you might need to turn off CTCSS or DCS to be able to talk to other people who don't have these tools.

7. Legal Considerations:
   - Learn about the rules in your area about using CTCSS and DCS tones to make sure you're following the law.

Group Communication Strategies

Setting Up and Managing Radio Groups

1. Channel Programming:
   - ❖ On your Baofeng UV-5R, press the **Menu** button.
   - ❖ Press the up and down button keys to get to the Memory option, and then press the Menu key.
   - ❖ Pick out the memory channel you want to save, and then press the Menu button.
   - ❖ Use the keyboard to enter the frequency for that channel, and then press Menu to confirm.
   - ❖ Do this again for every channel you want to add to your groups.

2. Assigning Channels to Groups:
   - Put channels into different groups based on what you need to do. As an example:
     - o **Group 1:** Emergency channels
     - o **Group 2:** Channels related to work
     - o **Group 3:** Personal communication channels
   - To do this, the program channels one at a time into specific memory spots or numbers based on the groups you've set up.

3. Labeling and Identifying Groups:
   - Give these groups names or labels so they are easy to find on the screen.
   - Go to the Memory options, pick the channel you want to name and use the keys to enter a name.

4. Dual Watch/Dual Standby Configuration (Optional):

- The Baofeng UV-5R radios can do **Dual Watch and Dual Standby**.
- Use these features to keep an eye on more than one show at the same time. Set up these modes to control how different groups or frequencies talk to each other.

5. Scanning Channels and Setting Priorities:
- Use the monitoring tool to keep an eye on or give more attention to certain channels in a group.
- Use the tracking and prioritization options to keep an eye on the most important channels in each group.

6. Testing and Making Changes:
- Make sure the features work by moving between channels and groups to make sure they are set up properly and are easy to find.

Selective Calling

Some channels can get busy with all the messages being sent when a lot of people use them to talk. To handle all of these texts, there are ways to block out extra or unwanted ones. When using a two-way radio, you can choose between two types of calls: group talking and individual calling. When you use group calling, you can talk to a lot of people at once. Everyone in the group can hear you when you talk because every radio is set up the same way. Pagers, which are used for individual calling, are like having a secret chat. You can only talk to a certain radio by giving a code that matches theirs. It's like calling someone on your cell phone.

The Baofeng UV-82 radio has three different ways to do group calling:

1. CTCSS
2. DCS
3. Tone-burst (1750Hz)

But, the Baofeng UV-82 doesn't have a feature for individual calling right now. When you use group talking, it's important to remember that other people can still hear your texts. These tools only help keep out information that you don't want. No matter what talking options you use, people who aren't using similar filters will still be able to hear your calls.

Conducting Group Calls and Announcements

Calling a User Group

❖ **Choose the Channel**: Turn the channel button to pick the channel that goes with the current talk group's name or ID.

❖ **Hold the Device and Place It**: Hold the walkie-talkie straight up and away from your mouth by about 1 to 2 inches (2.5 to 5 cm).

❖ **Press the PTT Key**: To begin the call, press the **PTT** key. The LED light goes red, and the talk group alias or ID and the group call sign are shown on the screen.

❖ **Let Go to Listen**: To listen, let go of the [**PTT**] key. The LED light goes green when the other radio replies. The screen shows the call group alias or ID, the group call sign, and the alias or ID of the other walkie-talkie.

❖ **Channel Idle Tone (if enabled):** If the **Channel Idle Tone** is turned on, you'll hear a short beep when the other radio stops sending. This lets you know that the channel is free for your answer. To answer, press the **PTT** key. A call stops after a certain amount of time if no one speaks.

❖ **Going back to the Normal Screen**: When the call is over, the walkie-talkie screen goes back to how it was before. You can also use your address book to start a group call.

CHAPTER 8
CTCSS

You can set a custom tone frequency for this system, which works like a "key" to block out unwanted messages on the same channel.

Here are the steps you need to take to set up CTCSS on your Baofeng radio:

1. To get to the menu, press the [**MENU**] button.
2. To get to **Receiver CTCSS (R-CTCS)**, press [**1**] [**1**] on the number keyboard.
3. To choose, press the [**MENU**] button.
4. Use the numeric keypad to enter the CTCSS sub-tone frequency you want in hertz.
5. To confirm and save your choice, press the [**MENU**] button.
6. To get to **Transmitter CTCSS (T-CTCS)**, press [**1**] [**3**] on the numeric keypad.
7. To choose, press the [**MENU**] button.
8. Use the numeric keypad to enter the CTCSS sub-tone frequency you want in hertz.
9. To confirm and save your choice, press the [**MENU**] button.
10. To leave the menu, press the [**EXIT**] button.

To turn off CTCSS and go back to normal channel use:

❖ Do the same steps as above, but in steps 4 and 8, enter [0] instead of a sub-tone frequency.
❖ To make sure you want to save your choice, press the [MENU] button.
❖ To leave the menu, press the [EXIT] button.

The CTCSS Table In the Technical Specifications section near the end of your radio's manual has a full list of all the available CTCSS codes and the sub-tone frequencies that go with them.

Emergency Communication Procedures

How to Enter a Basic Emergency Frequency:

Almost right away, you can use the UV-5R. It's very easy to use—just turn it on and choose a basic frequency or channel to send and receive. It's likely your first time using the radio, so let's start with the basics.

Reset/ "Zero Out" The Radio

Make sure the battery pack is snapped to the back of the transceiver before setting up the radio. Put the antenna on the antenna post and tighten it up. To turn on the radio, turn the volume knob counterclockwise. The radio will beep twice, and then a voice will say "Frequency Mode" or "Channel Mode." In this order.

TIP: To make sure there aren't any pre-programmed settings that could get in the way of emergency communications, you should "zero out" (reset) the radio to its original settings:

- ❖ Hit the **MENU** key.
- ❖ To get to menu option 40, use the up and down arrows on the keypad.
- ❖ Choose "**ALL**" by pressing the **MENU** button again.
- ❖ Select "**SOURCE**?" by pressing the **MENU** button a third time."
- ❖ To turn the radio back on, press **MENU** four times.

Select your preferred language

The radio will reset and default to a Chinese voice. To select your preferred language:

- ❖ Hit the **MENU** key.
- ❖ Find option **14** on the menu.
- ❖ To change the language, press the MENU button again.
- ❖ Find "**ENG**" for English (or your chosen language) with the arrows.
- ❖ To make sure you want to use that language, press MENU again.
- ❖ Exit.

Use the UV-5R as an FM radio

You can use the UV-5R as a simple FM radio to listen to your favorite stations. This is helpful during disasters when local radio stations send out emergency broadcasts and information. Just press the orange "CALL" button on the radio to turn on FM mode. By repeatedly pressing the */"SCAN" key, you can access every station.

Enter, save, and use an emergency frequency

By pressing the right numbers on the keypad, you can set a frequency and begin sending and receiving. If you type in 162.400, for example, you will be taken to the NOAA weather broadcast. Simply typing in 151.940 will connect you to the main national emergency line. That way, we won't have to remember all the numbers for each frequency we use in case of an emergency.

This is how you save a frequency and make a new channel:

1. To put the radio into Frequency (VFO) Mode, press **VFO/MR**.
2. To choose the top frequency, press the A/B button. Take note of the arrow to the left of the frequency on the screen. This shows that you have made your choice. The top frequency must be used for all programming.
3. Turn off **TDR/Dual Standby** (it should already be off, but make sure).
   o Press the menu button.
   o Press 7.
   o Press **MENU** to choose the menu item.
   o Choose "OFF" with the up and down arrows.
   o Press Menu to make sure.
   o Exit.
4. Use the keypad to enter the frequency you want to save.
5. Hit the **MENU** key.
6. Go to the 27th option.
7. Press MENU again to get to the channel list.
8. Pressing the up and down arrows will let you choose the channel you want (000 to 127). Start with channel 1, and then move on to channel 2, and so on. If there is "CH-" in front of a channel number, that channel already has a frequency saved.
9. To save the frequency to the chosen channel, press the MENU key.

10. Exit.

Now you can choose the saved frequency by pressing the up and down arrows and VFO/MR to switch to Channel Mode. The radio will play each channel from the list of saved frequencies. The screen will show you two saved frequencies and the channel where each frequency is saved while you are in Channel Mode.

Delete a saved frequency/channel

Deleting a frequency or channel is even easier:

- ❖ Hit the **MENU** key.
- ❖ Press 27/28 option.

- ❖ To change channels, press the **MENU** key.
- ❖ Pick out the channel or frequency that you want to get rid of.

❖ To get rid of it, press **MENU** again.

❖ Exit.

Search for active frequencies and transmissions

You could be in a disaster area where you don't know what frequencies or channels to use.

In this case, you can still use the UV-5R to listen in on emergency calls by scanning the airwaves:

- To make sure the radio is in **Frequency Mode**, press VFO/MR.
- Hold down the "**SCAN**" key for a while.
- If the radio hears a transmission, it will stop scanning through frequencies quickly.
- Press **MENU** to change how many frequencies the radio jumps between each scan.
- Select "**STEP**" as the first option.
- To choose a step, press the **MENU** key.
- You can change the size of the step by using the up and down arrows.
- In terms of speed and depth, the lowest step (2.5K) is the worst. 50K is the fastest frequency search, but it doesn't look very deep.

Change the radio's operating band (VHF or UHF)

The Baofeng has two frequency bands: Very High Frequency and Ultra High Frequency. At any given time, only one of the two bands can be used and watched.

To change bands:

- Hit the **MENU** key.
- Find option **33**: "**BAND**."
- If you want to change the band, press MENU again.
- You can choose VHF or UHF with the up and down arrows.
- To be sure, press MENU again.
- Exit.

NOTE: The VHF band is used by many police, EMS, government, and rescue operations for emergency radio frequencies.

CTCSS and DCS ("private line" or PL communications)

At times, more than one operator will use the same radio frequency to send and receive signals. This is very likely to happen during a disaster. To make sure communications stay stable, many first responders, command centers, and rescue operations will stick to one frequency. Tonal frequencies, on the other hand, are used to keep all the operators on the same frequency away from each other so that they don't share transmissions and make the airwaves too crowded. The Continuous Tone Coded Squelch System (CTCSS) and Digital Code Squelch (DCS) are the names of these tone frequency systems. The main difference between the two systems is that DCS is digital.

- Seventy hertz (Hz) are used to measure the 50 universal tones in CTCSS.
- DCS has 105 universal tones that are all numbered from A to Z (D023N).

To have a better understanding of this, consider a wireless phone and the cellular network it operates on. The CTCSS or DCS tone is like the phone number, and the frequency itself is like the network. To send or receive, you need to be on the network and to talk, you need to "**dial**" the right number (the CTCSS or DCS). Tones that work with both systems are already programmed into the UV-5R. To send, you need to know what tone the frequency in question is using.

How to Program CTCSS and DCS to a Frequency/Channel

To program a CTCSS or DCS tone into a frequency that requires it (and save it to a channel):

1. Press VFO/MR and put the radio in Channel Mode.
2. Ensure you're on Channel A by pressing A/B.
3. Type in the frequency you want to save.
4. Press **MENU**.
5. Navigate to option 10 and 12 to set a transmitting and receiving DCS tone.
6. Navigate to option 11 and 13 and select a transmitting and receiving CTCSS tone.
7. Press MENU again to select the appropriate option.
8. Use the up and down arrow keys to select the appropriate DCS or CTCSS transmitting and receiving tones. In options 10 and 12, or 11 and 13.
9. Press MENU to confirm your selection for each.

10. Navigate to option 27 to store the frequency and the transmitting and receiving DCS or CTCSS tones to a channel.
11. Exit.

Now, the frequency you just saved should be on the display with either "DCS" or "CT" to the left.

CHAPTER 9

INTERFERENCE AVOIDANCE

Identifying and Resolving Interference

- **Connect via USB**: To connect your radio to your computer, use a USB connection.
- **Open the CHIRP software**: On your computer, open the CHIRP software.
- To use the radio, turn it on and turn the volume up (100%).
- **Download Radio Settings**: Use the software to connect your radio to your computer via USB and download the current settings.
- **Find Squelch Settings**: Use the software's tabs to find the part where you can change the squelch settings. You could find it in a certain section, like "**Settings**" or "**Service Settings.**"
- **Change the Squelch Ranges**: Type in the new squelch ranges that are needed for both VHF and UHF frequencies where they are asked.
- **Upload Changes to Radio**: Use the software to connect your radio to your computer via USB and send the new settings.
- **Test Function**: Now that the mute settings have been changed, make sure they work right by testing them.
- **Repeating for Best Results**: If you need to, repeat the steps to fine-tune the settings for the best results.

Monitoring for Local Regulations

Monitoring FRS/GMRS Traffic with a UV-5R

First some definitions courtesy of the FCC. The Family Radio Service (FRS) is a private voice and data service that lets families and groups talk and send messages over short distances. People most often use FRS channels to talk back and forth over short distances using small hand-held radios that look like walkie-talkies. These are the 22 channels that the FRS is allowed to use between 462 MHz and 467 MHz. They are all shared with the GMRS. There are some differences between FRS and GMRS. GMRS uses the same frequencies as FRS, but it has extra features like being able to send data with more power and "short data messaging applications including text messaging and GPS location information." You will need to pay 70 dollars for a license, though. These are the frequencies that come with walkie-talkies that you can buy at a big box store. But most of the time, those walkie-talkies only show the channels and not the frequencies.

One last thing and this is a big one, before we start writing.

It is against the law to use a Part 90 device, like a Baofeng UV-5R or one of its siblings, to broadcast on the FRS or GMRS frequencies. That being said, I haven't found any examples of the FCC doing this, but it's something to be aware of. To follow the law, you must have Part 95-approved radio, such as the Baofeng GMRS-V1, or one of the many ready-made options. You can also use things like channel bandwidth without a license, but I'm not going to talk about them since you shouldn't be sending on a UV-5R. We are going to use CHIRP to set the UV-5R with these frequencies, just like we did before.

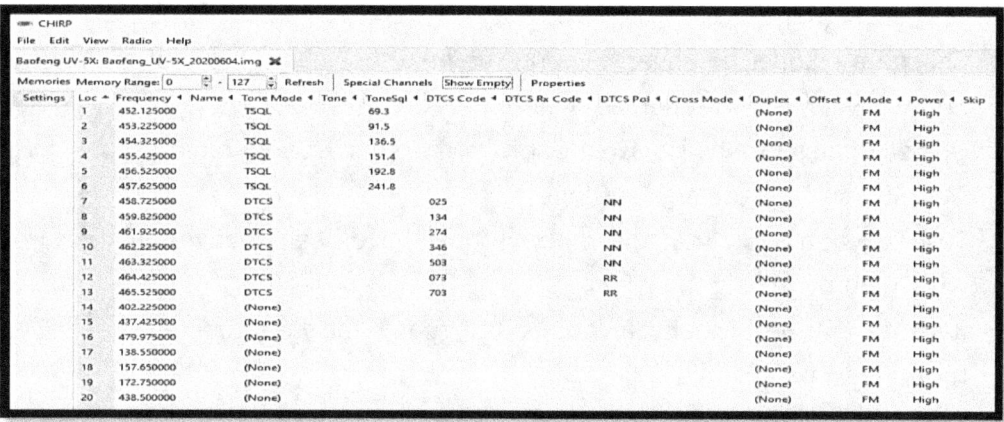

Now pick it all out and wipe it! To get to the next step, go to **File > Open Stock Config > US FRS and GMRS Channels.**

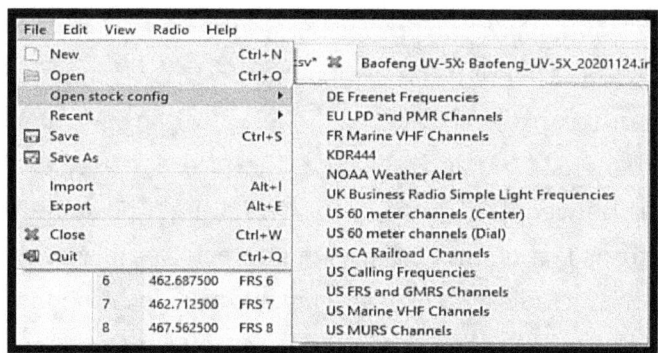

When it's open, it will look like this.

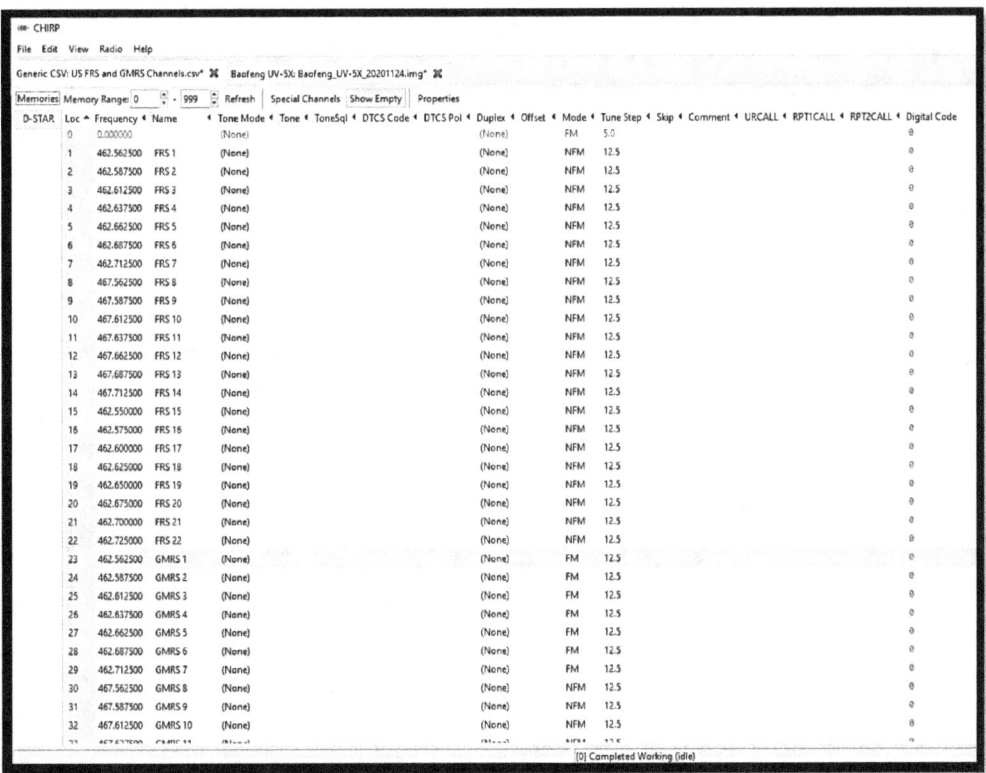

You can choose channels FRS 1–22 from this page, copy them, and then put them into your UV-5R file. The UV-5r doesn't care about the Tune Step column, so don't worry about it. You may have also noticed that some of the frequencies are marked GMRS 1–22. Technically, you don't need to copy these in because they are the same frequencies. However, for fifteen of the GMRS channels, the buffer size is different. If you want to be sure you hear everything, you can also paste them in. This doesn't change what you'll hear, but it's optional.

*On the UV-5R at least. I had some trouble with my Radioddity GD-77 when I changed the bandwidth. It started sending out junk through the speaker.

This is what you'll have when everything is done.

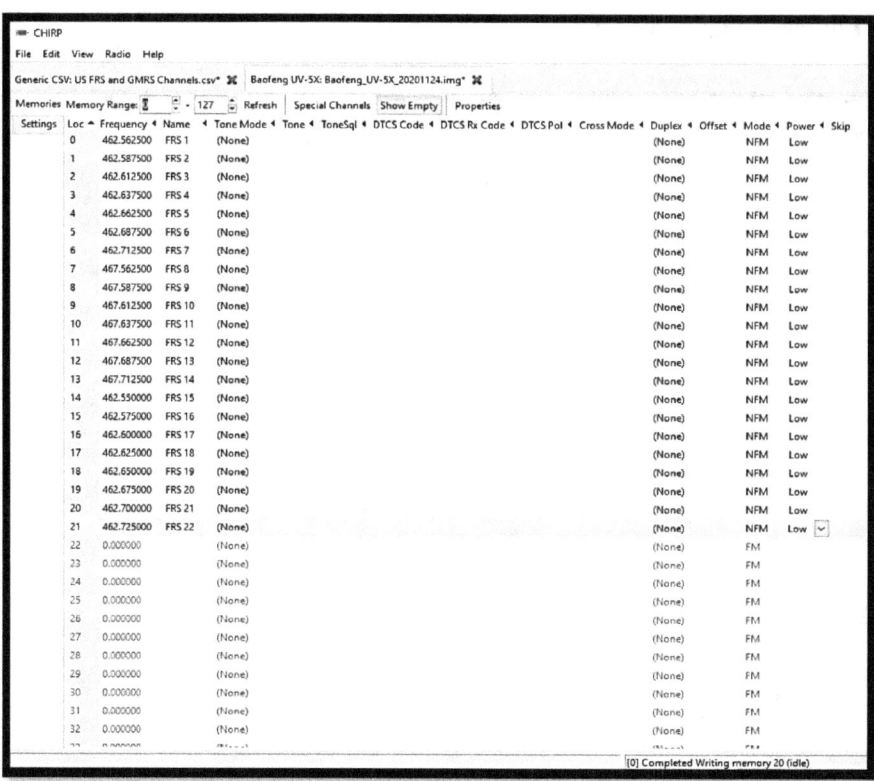

I think you should set all Power fields to Low so that if you send something by mistake, you won't be breaking FRS's power limits. Anyway, I think you should not send anything, so I would just turn off VHF/UHF TX in the Other Settings Menu.

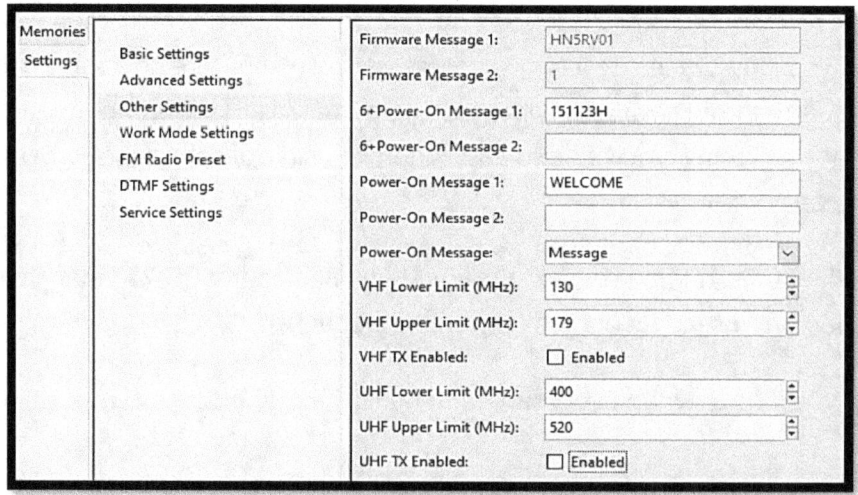

Also, these are all simplex channels, so don't worry about the duplex field; it can stay at zero. Finally, save your image file, connect your radio, and share it. Then press VFO/MR on the radio and you'll be with A and B on the list of favorites. Holding down ***** will start the scan. The A and B don't matter for screening, but they might be helpful if you want to keep an eye on two frequencies that you know have business on them.

You might be wondering why I didn't tell you to add any tones.

Well, they don't matter at all for tracking. When your tone is set to None/CSQ, you can hear all transmissions on a frequency, even if the sender is using a tone or "privacy code." They don't protect your privacy at all and only let multiple people use the same frequency. If you want to know what tone is being used, you can "**capture**" it, but the UV-5R makes it hard to do. On a radio, encryption is the only way to maintain privacy, which is not present here. To sum up, anyone can hear anything that is sent on FRS/GMRS (or any other channel that isn't encrypted). You might be curious about what you can hear and how far away it is now that you have set your UV-5R to the FRS/GMRS frequencies. How you answer that question depends on where you are, what kind of antenna you have, and what kind of radios is being used. You might be able to hear someone using an FRS/GMRS radio within a few blocks of you if you're at a protest in a city, though it depends on how strong their signal is.

Per the FCC

Between channels 8 and 14, an FRS device's range is usually less than half a mile. On channels 1 through 7, and between 15 and 22, based on the conditions, interactions can be farther away. Of course, what you hear will depend on the people who are using the radios, provided anyone is. I thought it was most interesting to hear kids talking to each other from their homes while I was trying one of my antennas. Your mileage may be different. Just a quick note about GMRS. If someone is using a radio that can handle GMRS and send with more power, their signal will probably go farther. On the other hand, they have to send their name at least once every 10 minutes and at the start and end of each broadcast. All callsigns are unique and belong to a single person.

If you have a GMRS license and are using a radio that can use GMRS, don't forget to send your name! Now, someone with a UV-5R might not be using the FRS/GMRS frequencies. If they have a license, they could be using MURS or the 2-meter/70-centimeter public radio bands. You can find MURS under the US MURS Channels in the **File -> Open Stock Config** menu. It's pretty simple to set up. There you can also find the National Calling Frequencies (Ham Simplex) for 2 meters and 70 centimeters. The ARRL band plans say that frequencies between 146.40 and 146.58 MHz and 147.42 and 147.57 MHz are single. If you want to get angry, you could look up the band plans for your local radio club and program in their simplex frequencies as well. That being said, it's not likely that people who aren't approved will use these frequencies. Now you can use a UV-5R or another CHIRP-compatible radio to listen in on the FRS and GMRS frequencies. Keep an eye on this space for a link to the image file that has all the frequencies in it so that you can easily load it onto your UV-5R.

Scanning

There is an inbuilt reader in the Baofeng UV-82 that works on both VHF and UHF bands. When the scanner is in Frequency (VFO) mode, it changes frequencies based on the frequency step that was set. When it is in Channel (MR) mode, on the other hand, it looks through the saved channels. By hitting and keeping the [SCAN] button for about two seconds, you can turn on the scanner. Any button can be pressed to get out of the scanning mode.

There are three different ways to use the scanner: Time, Carrier, and Search. To change the reader setting, do the following:

How to Change the Scanner Mode

- ❖ To get to the menu, press the [MENU] button.
- ❖ To get to the reader mode, use the number keyboard to enter [1] [8].
- ❖ Press the [MENU] button to pick this choice.
- ❖ Use the [UP] and [DOWN] keys to choose the reading method you want to use.
- ❖ Press the [MENU] button to make sure you want to save your choice.
- ❖ Press the [EXIT] button to leave the page.

Let's dive into the different scanning modes:

- ❖ **Time Operation (TO):** When the reader finds a signal in Time Operation mode, it stops. It starts searching again after a set amount of time.
- ❖ **Carrier Operation (CO):** When the scanner is in this mode, it stops searching when it finds a signal and only starts again when the signal is lost.

❖ **Search Operation (SE):** When the scanner finds a signal in this mode, called "Search Operation," it stops.

To scan again in any of these modes, hold down the [SCAN] button for a while. These modes have different features that are designed to meet different needs. This lets users change how the scanner works based on their tastes and the needs of their communication jobs.

Tone Scanning

You can use the following steps on your Baofeng radio to look for CTCSS (Continuous Tone-Coded Squelch System) tones or DCS (Digital-Coded Squelch) codes on busy frequencies when it is in frequency mode:

Procedure for Tone Scanning

∞ To get to the menu, press the [MENU] button.
∞ Enter one of the following numbers using the number pad: a. Type **[1] [0]** to start a search for DCS codes. You can start looking for CTCSS sub-tones by pressing **[1]** **[1]**.
∞ Press the [**MENU**] button to make sure you want to make the choice.
∞ Press the [**SCAN**] button for a short time to begin scanning.
∞ The monitor will show "**CT" or "DCS**" flashing as the radio starts to scan. This means the radio can hear the tone or code that has been given. While listening, the radio will beep and stop flashing to let you know that it has found a tone or code.
∞ Press the [**SCAN**] button to confirm the tone or code you found.
∞ Press the [**EXIT**] button to leave the page.

You can actively look for and spot specific DCS codes or CTCSS tones that are being used on current frequencies with this method. By doing these steps, you can make good use of your Baofeng radio's ability to search for and pick up these signaling codes. This can help you keep an eye on things or set up your radio so that it can talk to other people using these tone or code settings.

Dual Watch / Dual Reception

The Baofeng UV-82 radio has a unique function called Dual Watch that lets users listen to two channels at the same time, even though the radio only has one antenna. This feature lets the device automatically switch between two frequencies every so often, which make the conversation more flexible. But it's important to know that some functions are turned off when the phone is in Dual Watch mode so that the feature can work.

Functions Disabled in Dual Watch Mode

- Reverse function
- Usage of the [POUND] button to switch between high and low transmit powers in channel mode
- Saving of duplex channels

To activate or deactivate the Dual Watch mode and manage its settings, follow these steps:

Procedure: Enabling or Disabling Dual Watch Mode:

- ❖ To get to the menu, press the [MENU] button.
- ❖ To get to Dual Watch, press [7] on the number keyboard.

- ❖ Press the [MENU] button to pick this option.
- ❖ To turn on or off the Dual Watch mode, press and hold the [UP] and [DOWN] keys.

- ❖ Press [**MENU**] to make sure you want to make your pick.
- ❖ Press the [**EXIT**] button to leave the page.

When the radio is in Dual Watch mode, it may send on channel A or B, depending on which channel becomes active first. That being said, this could be a problem if the current frequency doesn't let the gadget send. To fix this, you can choose to lock the radio to either channel A or channel B. Here's how to activate this feature:

Procedure: Locking the Dual Watch Transmit Channel

- Press the [**MENU**] button to get to the menu.
- To get to **TDR-AB**, press [**3**] [**4**] on the number keyboard.
- Press the [**MENU**] button to pick this choice.
- To switch between screens A and B, press the [UP] and [DOWN] keys.
- Press the [**MENU**] button to make sure you want to make the choice.
- Press the [**EXIT**] button to leave the page.

For a short time, you can bypass the lock without turning off the menu, and press the [**PTT-A] or [PTT-B**] button before pressing the [**PTT**] button.

These features give users more control over the Dual Watch mode, letting them set their broadcast settings and making sure that communication on the Baofeng UV-82 radio runs smoothly and quickly.

CHAPTER 10
ADVANCED FEATURES AND FUNCTIONS

Dual-Band Operations

A radio can send and receive messages on two different frequency bands if it can operate in dual-band mode. A lot of the time, Baofeng radios is made to work on both the Very High Frequency (VHF) and Ultra High Frequency (UHF) bands. People can talk on different frequencies within these two bands because these radios can switch between them. The range of VHF frequencies is 136 to 174 megahertz (MHz), and the range of UHF frequencies is 400 to 520 MHz. There are pros and cons to each frequency band. When the area is open, VHF signals tend to move farther and can get through buildings and plants better than UHF signals.

This means that VHF can be used outside or in places with fewer obstacles. On the other hand, UHF signals work better in cities or places with lots of obstacles because they can get through buildings and other structures more easily. UHF broadcasts also work better indoors and in places with a lot of people. Users can switch between VHF and UHF frequencies on Baofeng radios that have dual-band abilities based on their needs and their surroundings. This makes it possible to talk over longer and shorter distances and on different types of terrain. This makes these radios useful for a wide range of situations, including outdoor sports, emergency contact, amateur radio (ham radio) use, and more. To sum up, Baofeng radios that work on two bands can send and receive messages on both VHF and UHF frequencies. This function lets users pick the frequency band that works best for them, whether they are outside, in a city, or inside.

Exploring Dual-Band Capabilities

- **Frequency Range**: Compared to single-band radios, Baofeng radios with dual bands can handle a bigger variety of frequencies. Most of the time, they can work on VHF frequencies between 136 and 174 MHz and UHF frequencies between 400 and 520 MHz.
- **Communication Options**: Users can communicate in a variety of ways when they can switch between VHF and UHF bands. VHF is good for long-distance contact in

open areas, while UHF is better for cities or places with lots of hurdles because it can go through them more easily.

- **Flexibility**: Dual-band radios, like those made by Baofeng, can be used in some situations. You can move to the VHF band, for example, in an emergency where you might need to talk over long distances and open land. The UHF band might work better in busy towns or places where things can get in the way of signal delivery.

- **Amateur Radio (Ham Radio) Use**: People who are interested in amateur radio (Ham Radio) often choose dual-band radios because they give them access to more frequencies. This can be very useful for communicating within certain ham radio bands and for sending sound, data, or digital signals.

- **Channels and Programming**: Most Baofeng radios have customizable channels that let users save and get to different frequencies in the VHF and UHF bands. This function makes it easy to switch between frequencies that are used by different groups or for different reasons.

- **Interoperability**: Some dual-band radios can receive on one band and send the signal again on the other band at the same time. This is called cross-band repeat. This can help send messages between different radio bands or make contact farther away.

- **Enhanced Features**: Dual-band radios often have extra features like dual-watch or dual-receive that let users listen to or send signals on two different frequencies at the same time.

- **Learning Curve**: For people who are just starting, the dual-band features may seem too much at first because there are so many options and settings. Users can use these features well for conversation, though, after some time and getting used to them.

Utilizing Cross-Band Repeat Function

Cross Band Mini-Repeater

The project's goal is to set up two transceivers from the Baofeng/Pofung series so that they can work as a remote base, a transmitter for Fox Hunt, and a mobile cross-band one-way repeater.

Here is a thorough guide on how to set up these features with all the links and requirements you need:

Requirements: For this setup, you'll need

- A pair of Baofeng/Pofung series transceivers
- A 2.5mm/3.5mm audio cable

Connection

The audio cable's 2.5 mm end should be plugged into the first Baofeng transceiver's top jack. As the receiver, this unit will do its job. The 3.5 mm end of the same audio wire should be plugged into the second Baofeng transceiver's lower jack. As the broadcaster, this unit will do its job.

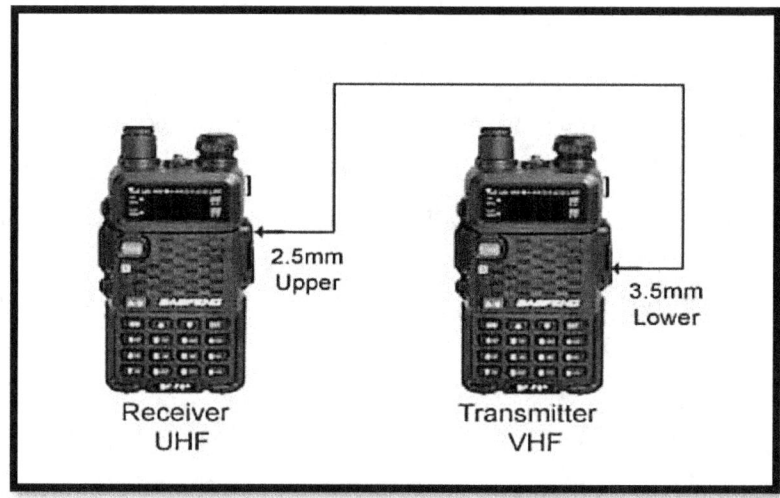

1. VOX Activation:
   - ❖ *Voice-operated exchange (VOX)* must be turned on in the sending (TX) unit. You can change the VOX Level to suit your tastes:
     - ○ **VOX Level 1:** The transmitter (TX) goes off about two seconds after the receiver (RX) silence stops.
     - ○ At **VOX Level 10**, the TX level drops as soon as the RX delay closes.
   - ❖ Turn up the volume on the radio that is receiving (RX) until it is comfortable.

Added Considerations for Field Operation

Power Conservation

- To save power, turn off functions that aren't needed, such as display lighting (ABR) and Roger Beep.
- For better results, use high-quality antennas like J-Poles.
- To cut down on crosstalk, put the RX antenna at least 15 feet above the TX antenna.

Separation and Isolation

Frequency Selection

- To keep disturbance to a minimum, set one radio to UHF and the other to VHF.
- Pick frequencies that are as far apart as possible to keep signals from getting mixed up. To stop harmonics, stay away from frequency multiples.
- As an example, a sound at 147.0 MHz can cause problems at 441.0 MHz with its third harmonic.

2. Antenna Separation

   ❖ Make the space between antennas bigger so that radios don't interfere with each other as much.
   ❖ Less disturbance happens when antennas are farther apart.

Duplexers for Single Antenna Use

Cross-Band Operation (VHF/UHF) using a Single Antenna

- For cross-band transmission, you could use a small dual-band duplexer like the Opek DU-500 or the MFJ 916B to run both radios off of a single antenna.
- Both radios can use the same antenna with a dual-band duplexer because VHF and UHF waves don't interfere with each other.

Operation on the Same Band

- A single-band duplexer is needed for both radios that work on the same band. But these are usually more expensive, and this simple job might not be a good fit for them.

Note on Duplexer Usage

- Duplexers are usually made to be used inside or on the go, so they might not work well outside, especially when it rains.
- Don't use duplexers outside when it's raining or snowing to make sure they work right and last a long time.

These settings and things to think about make Baofeng radios more useful, cut down on disturbance and work better when used for different tasks, such as cross-band activities and saving power in the field.

Digital Modes and Encryption

Understanding Digital Modes (DMR, D-Star, etc.)

Digital Modes techniques are used to send information over radio waves in digital forms like DMR (Digital Mobile Radio), D-Star, and others. For efficient and clear contact over radio channels, each method has its own set of rules and procedures for encoding and interpreting data. Amateur radio fans like the Baofeng radios because they are affordable and can be used in a variety of situations. Depending on the type and software, these radios often have more than one mode, such as analog and digital.

Here's a brief overview of some common digital modes:

1. **DMR stands for "Digital Mobile Radio."** DMR is used a lot in both business and hobbyist radio. Time-division multiplexing lets two talks happen on the same channel, and it's known for being good at using bandwidth efficiently. Talk groups are a common way for DMR radios to organize conversations.
2. **D-Star (Digital Smart Technologies for Amateur Radio):"** is another digital talk and data system used in amateur radio. It was created by the *Japan Amateur Radio League (JARL).* It has features like digital talk, data, and even GPS data transmitted in the stream.
3. **Software called System Fusion (C4FM)** was created by Yaesu. It is a digital communication system that blends digital and traditional speech. It has abilities like sending data and recording sound.

It's important to make sure that the Baofeng radio you want to use is compatible with the digital mode you want to use. To make digital mode work, you might also need to add more devices or update the software.

Here are some steps to utilize digital modes on a Baofeng radio:

- ❖ **Check Compatibility**: Make sure that the Baofeng radio type you want to use works with the digital mode you want to use (DMR, D-Star, etc.). Check to see if the Baofeng type you want to use supports digital modes before you try to use it.
- ❖ **Programming**: Set up the radio with the frequencies, talk groups (if you're using DMR), and any other settings it needs for the digital mode you want to use. To set up your radio, you may need software and connections that work with it.
- ❖ **Firmware Updates**: On some radios, updates to the firmware can add new features or make digital mode work. Check to see if any software changes for your Baofeng model could make digital mode work.
- ❖ **Use the Right Accessories**: If you're using a digital mode, you may need a digital booster or hotspot that works with that mode.

Remember that if you want to use digital modes, you usually need to know a lot about the methods and settings that go with each mode. Also, it is very important to follow the rules and get the right licenses for radio contact, especially when using amateur radio.

Implementing Encryption for Secure Communication

Encryption can make it hard to use radios for private contact, especially in amateur radio or public bands, because of legal limits in some places and the way encryption works. Encryption on some radio frequencies is illegal or limited in many countries to make sure that communications are clear and easy to get to. But private contact might be okay in certain approved bands or for certain uses, like government or public safety.

Here are some things to think about if you want to use encryption for radio communication:

1. **Regulations and Legalities**: Before you use encryption, make sure you're following the rules and laws in your area about radio. In some places, it is illegal to use private communication on certain bands or without certain licenses. Find out about the rules that apply to where you live and the frequencies you want to use.

2. **Licensed Bands or Modes**: Encryption can be used with some approved radio services or modes. In the permitted spectrum of some government, public safety, or private radio services, for example, there are ways to communicate securely. You might be able to use encryption if you have the right license for these bands.

3. **Specialized Systems**: Some business systems, like DMR, P25, or TETRA, can secure data. But to use these systems, you usually need special tools and rights that aren't easy to find or aren't legal to use on amateur radio or public bands.

4. **Open-Source Encryption**: Open-source encryption methods could be used in some situations as long as they don't break the law. These might not offer the best level of security, but they might give you some privacy. Some apps, like WAVE and FreeDV, offer some amount of security for digital voice calls over ham radio bands.

5. **Privacy Measures**: Use of codes, selective calls, or lesser-known modes could add an extra layer of privacy to your conversation without actually hiding the information. Full encryption might not be possible.

6. **Respecting Security and Privacy**: Make sure that the encryption or privacy measures you're using don't get in the way of emergency contacts or other important services. Always put the safety and security of public services ahead of your contact.

GPS and APRS Integration

To add GPS and APRS (Automatic Packet Reporting System) features to a Baofeng radio, you need to use tools that work with the radio and know how to set it up for APRS communication. Most Baofeng radios don't have GPS or APRS built in, so you'll need to use extra devices and links to join them.

Here's a basic guideline for integrating GPS and APRS functionality with a Baofeng radio:

❖ **GPS Unit**: You'll need a separate GPS device that can tell you where you are. A lot of the time, these GPS units has cables that connect to the radio, and they may have specific needs for working with your Baofeng model. Make sure it works with APRS and gives you the info output you need.

❖ **(APRS TNC)**: You'll need an APRS TNC to connect the Baofeng radio to the GPS and APRS system. The TNC handles the encoding and decoding of the APRS packets between the radio and the GPS. There are different types of TNCs, and you should pick one that works with your Baofeng radio and GPS unit.

❖ **Wiring and Connections**: Follow the directions to connect the GPS unit and APRS TNC to your Baofeng radio. Usually, this is done by using the right cords to connect the GPS unit to the TNC and then the TNC to the Baofeng radio.

❖ **Setting up**: Make sure that the APRS TNC and your Baofeng radio are both set up so that they can work together. Setting the right frequency, baud rate, and other parameters is needed to make sure that the GPS, TNC, and radio can talk to each other properly.

❖ **Testing**: Once everything is set up and linked, make sure the GPS data is being sent over APRS using your Baofeng radio by testing the setup. Make sure that your broadcasts follow the rules and frequencies set aside for APRS in your area.

Enabling and Using GPS Features

When turned on, this function gives you a lot of information, like your exact location and longitude, your height above sea level, the date and time, and more. It can also show your speed while you're walking or driving if you're moving.

To turn on the GPS device by hand:

1. To get to the main Menu, press the (**Menu**) button.
2. After scrolling down, choose **GPS**.
3. Pick whether the GPS is on or off.
4. To turn on the GPS device, select **GPS On**.

Alternatively, you can use the Customer Programming Software (CPS) to do this. Go to **Public → Optional Settings → GPS/Ranging** after starting the app. On the page that comes up, turn on both GPS and Get GPS Positioning. The GPS sensor will stay turned on by default after you reprogramme your radio.

Once the GPS is enabled, accessing the GPS information manually is straightforward:

❖ To get to the main Menu, press the (Menu) button.
❖ Pick out **GPS** by scrolling down.
❖ If you select **GPS Info**, the radio's LCD screen will show you full GPS data.

In addition, you can set one of the special keys to quickly show this information with a button press. For example, I've set up the PF3 key so that it shows GPS information when it's pressed briefly. Either the options or the CPS interface can be used to do this by hand.

Introduction to Automatic Packet Reporting System (APRS)

The Automatic Packet Reporting System (APRS) is a digital amateur radio communications system that is mostly used to send and receive tactical information in real-time, track a person's location, and send other types of data. Bob Bruninga, WB4APR, created it in the early 1980s. Since then, it has grown into a flexible method that amateur radio users all over the world use. Amateur radio frequencies are used by APRS, which lets people share and receive small data bits over the air. These packets can hold many types of data, including location data from GPS, weather reports, texts, tracking data from devices far away, and more.

Key components of APRS include

❖ **Position Reporting**: One of the most important parts of APRS is that it can keep track of where stations are at all times. Users can share their latitude, longitude, and altitude information by using GPS technology. This lets other users see where they are on digital maps almost instantly.

❖ **Instant messaging**: APRS allows short messaging, which lets managers send and receive text messages between stations. This function can be used to talk to each other during events, or situations, or just to share information between workers.

❖ **Reporting the weather**: APRS can send weather data from weather stations that have the right equipment. This includes things like temperature, humidity, air pressure, wind speed, and direction. It gives other users a specific weather report that they can view.

❖ **Object Tracking**: APRS lets users make and follow items as well as tracking sites. Objects can stand for many things, like cars, tools, or specific places of interest, making it possible to see them on maps.

❖ **Telemetry and Sensor Data**: APRS makes it possible for remote sensors to send telemetry data, which can be any kind of recorded data, like temperature, voltage, current, or anything else. This information can be sent automatically or at set times.

APRS works on different frequencies in the amateur radio bands, and the AX.25 system is usually used to encrypt the data. It uses a network of digipeaters (also called "digital repeaters") to send and receive messages, which makes contact more widespread.

The method has been used for public service announcements, emergency messages, search and rescue operations, tracking vehicles, reporting the weather, and some novel uses in the amateur radio community. APRS has always changed, adding new technologies and increasing its functions. This has made it a useful and flexible tool for amateur radio users who do a variety of tasks.

CHAPTER 11

LEGAL AND REGULATORY CONSIDERATIONS

Radio Licensing

Getting a GMRS License

Citizens Band Radio Service (CBRS), the Family Radio Service (FRS), the Multi-Use Radio Service (MURS), and the General Mobile Radio Service (GMRS) are some of the services that can be used for work or family interactions. GMRS might be the most useful of these four services.

For example, many GMRS channels are on the same frequencies as FRS channels. However, GMRS users can use more power than FRS users, set up repeaters to make their radios' range longer, and use some data apps like text messages and GPS position. But there is a cost to these talents. The **CBRS, FRS, and MURS** systems can be used without a license. However, to use a GMRS system, you need an FCC license, which costs $70 for ten years. To get a license, you have to be at least 18 years old and not work for a foreign government. However, a single license lets anyone in the family-run a GMRS station, no matter what age they are.

To get a **GMRS license**, you must first get an **FRN, or FCC Registration Number**. If you don't already have one, ask for one. To start, go to **CORES**, the Commission's Registration System, and click Registration. Select whether the FRN is for a person or a company, and check that the contact's address is in the United States. Then click "Continue." Fill out the form below with your information, and then click "**SUBMIT**." As soon as you finish the form, the CORES system will give you an FRN. The next thing you need to do is get your GMRS license. Type in your **FRN** and password in the **FCC License Manager**, then click **SUBMIT**.

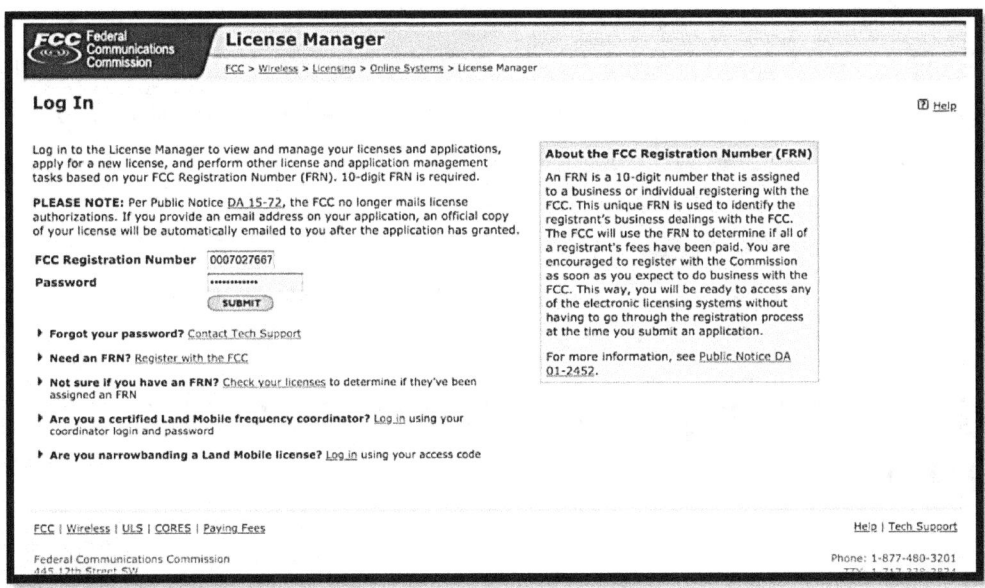

The next page will show your current licenses. I am a licensed radio amateur, so the screenshot below shows the status of my amateur radio license. To apply for a GMRS license, click on "**Apply for a new license.**"

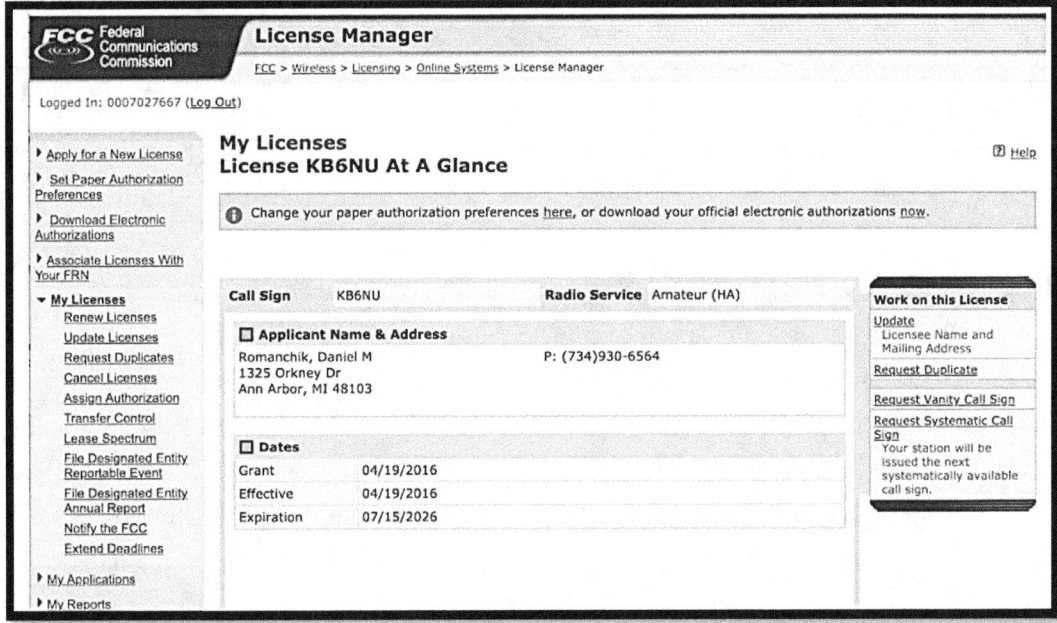

On the next page, use the drop-down menu to choose the last item, which is ZA – General Mobile Radio Service (GMRS). Then hit **CONTINUE**. Answer the questions asked by the candidate, then click "**Continue**." Type in your name and address on the next page, then click "**Continue**." If you have ever been linked to a crime, you will be asked on the next page. This is something new that all FCC license forms now have. Kindly answer the question and then click "**Continue**." A summary of your entry will be shown to you on the next page, as shown below. Click on the **EDIT** button next to the information you want to change.

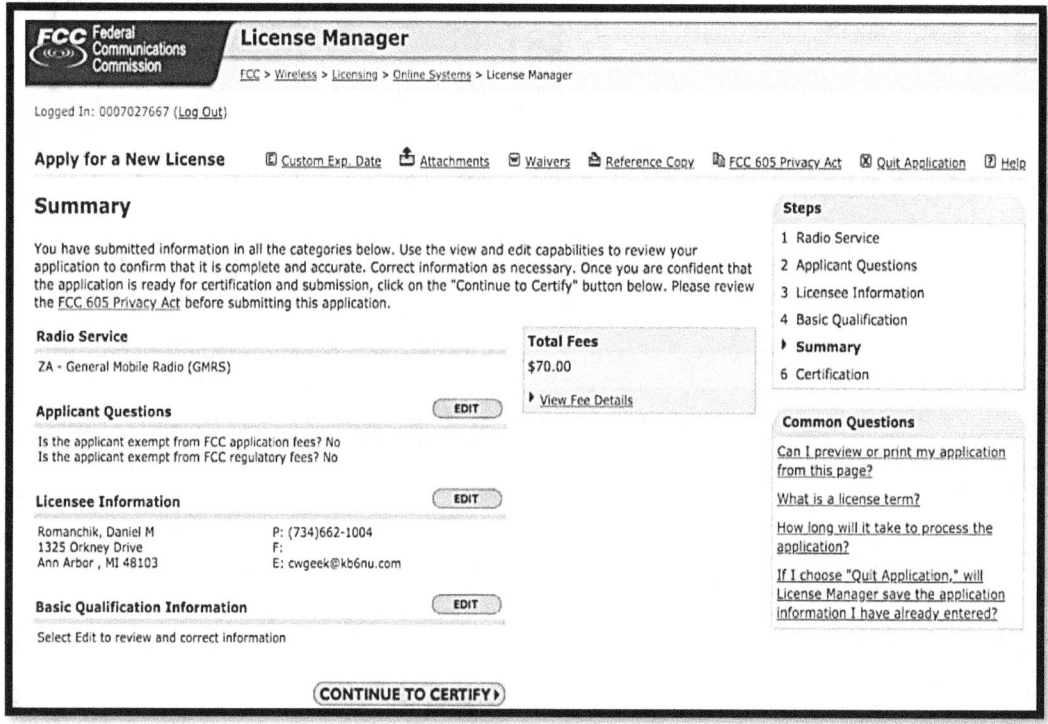

To get to the Certification page, click **CONTINUE TO CERTIFY**. You only have to confirm that you are qualified for a GMRS license on this page. Type in your name and a title, if you have one, and then click "**SUBMIT APPLICATION**." The next thing you'll see is a proof of your entry. To go ahead with payment, click "**CONTINUE FOR PAYMENT OPTIONS**." Use your FRN and password to log in again on the next screen. This will take you to a page like the one below, which has some payment options. I used a credit card to pay.

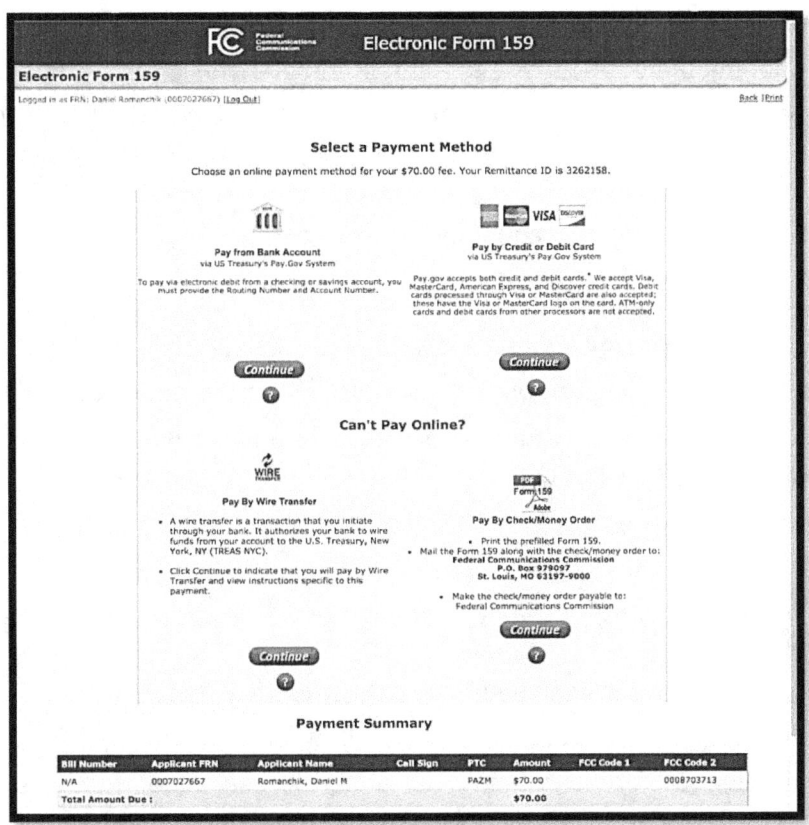

You'll need to enter your payment information and click the "**Continue with Plastic Card Payment**" button. After that, you'll be taken through some screens to make your payment. You will eventually get an email confirming that you paid with a credit card and letting you know that you have been given a GMRS license. Once you are back on the FCC website, log in and click the Download Electronic Authorizations link on the My Licenses page. This will let you download or print a copy of the license authorization.

Radio Frequency (RF) Exposure Limits

Radio Frequency (RF) exposure limits are the rules and guidelines that make sure people stay safe when they are around electromagnetic fields, especially those given off by radios like Baofeng radios. These limits were set to protect people from the possible health risks that come with being exposed to RF energy for a long time. In most countries, government organizations like the Federal Communications Commission (FCC) in the US, the International Commission on Non-Ionizing Radiation Protection (ICNIRP), and other similar groups around the world set and enforce limits on RF radiation. There are rules about RF radiation for Baofeng radios and other emitters. Following these rules makes sure that the radio's electromagnetic energy stays at safe levels when it is used in line with the official instructions.

The RF exposure limits take into account factors such as:

- ❖ **Specific Absorption Rate (SAR):** This is a way to measure how fast the body absorbs radio frequency energy when it is exposed to electromagnetic fields. It is generally given in watts per kilogram (W/kg) and helps figure out how much radiation is safe.
- ❖ **Range of frequencies:** The human body absorbs and carries different frequencies at different depths and rates. Often, different frequency bands have different radiation levels set by law.
- ❖ **Power Levels:** The amount of RF radiation is affected by how much power the radio emitter puts out. Most of the time, smaller power levels mean less contact.

Like most commercial transceivers, Baofeng radios should follow the limits on RF exposure set by government agencies. Users are told to follow the safety instructions given by the maker, like keeping the radio antenna at a certain distance from the body while broadcasting. It's important to remember that even though there have been worries about the possible health effects of RF exposure, so far science studies have not found strong proof that radio frequency emissions within legal limits are harmful to health. But it's smart to follow suggested safety measures to avoid needless contact.

To ensure compliance and safety:

- ❖ Follow the manufacturer's instructions on how to use and place the antenna correctly.
- ❖ Stay away from the receiver of the radio, especially when you're sending.
- ❖ Follow the rules and laws in your area about how much RF radiation you can be exposed to.

CHAPTER 12
RADIO ACCESSORIES AND ADD-ONS

Antenna Upgrades

A bad connection is the main thing that gets in the way of a good conversation. Plus, it makes things hard to understand between the workers. Once more, a good antenna is the best thing for Baofeng radio users to have with them. People all over the world love and use Baofeng radios because they are the best. You take your Baofeng radio with you on a hike or hunt and either don't have an antenna at all or only have a normal one that doesn't let you send and receive messages properly. After that, your adventure will feel like it's missing something. Can you guess what the thing is that will end your journey? The best Baofeng antenna is just a simple piece. But "**best**" means that there are many great antennas for Baofeng, each with its benefits. The people who do our studies are very good at understanding what people want. Because of this, they were able to find the best antennas for Baofeng, which have won over the hearts of many Baofeng radio users. So, let's find out more about them.

Why Should You Invest in Baofeng Antenna?

A lot of people believe that getting a Baofeng antenna should cost more. But I think Ham radio users should improve the Baofeng antenna. Do you know why?

Here are the reasons.

You should consider purchasing a Baofeng antenna for some reason.

- ❖ The first reason is that it is dependable.
- ❖ The second reason is that it is simple to put up a Baofeng antenna.
- ❖ The third reason is that its price is affordable for the average person.
- ❖ There is also the fact that its reception is extremely clear even when it is at a greater distance.
- ❖ Finally, the antenna is long-lasting.

Top list of Best Antenna for Top 10 model Baofeng Radio

- ❖ Authentic Genuine Nagoya NA-771-Best Antennas For baofeng UV-5r
- ❖ Bingfu Dual Band VHF UHF Ham Radio Antenna-Best Antennas For BaoFeng BF-F8HP
- ❖ 2 Pack-29 Inch Foldable/Tactical Radio Antenna-Best long-range antennas for Baofeng UV82
- ❖ 42.5-Inch Length ABBREE SMA-Female Dual Band Tactical Antenna-Best Antennas For baofenG-82
- ❖ BaoFeng bf-888s Antenna 10 X Original Antenna– Best Antennas For BaoFeng Bf-888s
- ❖ Baofeng Magnetic Car Vehicle Mounted Antenna-Best long-range antennas for baofeng UV-5r
- ❖ HYS SMA-Female Handheld Dual Band Antenna-Best antennas for baofeng uv 82hp
- ❖ BaoFeng SRH805S SMA-F Female Dual Band Antenna-UV-Best Antennas for baofeng 5R And BF-888s Radio
- ❖ Walkie Talkie Antenna 15.6-Inch Whip Dual Band UV-Best Antennas for baofeng UV-5RE
- ❖ Dual Band 136-174Mhz&400-520Mhz SMA-Female Antenna-Best short antenna for baofeng UV-5r

1. Authentic Genuine Nagoya NA-771 VHF/UHF Antenna

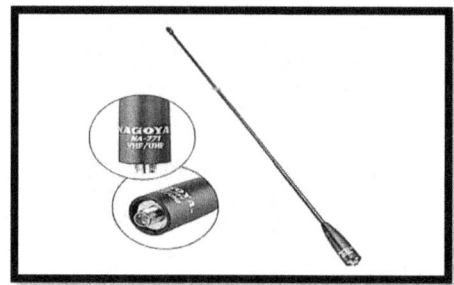

The Nagoya NA-771 15.6-inch whip VHF/UHF (144/430 MHz) antenna is now up and running. BTECH made this two-way radio antenna, which is one of the best for the Baofeng UV-5R. All ham radio users know and love the BTECH name. So, something from this brand is something to praise. The strong magnet on the base of the antenna keeps it straight on the roof of the car, even when the car is going fast. The reception is much clearer, and the material is strong and will last. Things like split plugs and long cords show how good the antenna is. NA-771 and NA-771R is pretty much the same thing. NA-771 works a little better than NA-771R, but it's not as flexible. It's scary that there are so many fake Nagoya antennas on the market. So, you need to be careful with this. Nagoya includes a small rubber screw or gap, which is a sign that the antenna is real. Its spring base works great for objects that hang low. It has good signal quality for medium-range reception and great signal quality for short-range reception. For long-distance transmission, it's a longer antenna that looks like a whip. But even when there is low pressure, the antenna caps can come off sometimes. Once more, the way it's made should be better.

Key features

- It works well on all bands. It also works with most Btech and Baofeng radios, such as the UV-5R, BF-F8HP, and UV-82HP.
- A better signal from faraway places with it. You can also enjoy its range of capacities.
- The most power it can handle is 10w. It works with frequencies between 136 and 174 MHz and 400 and 520 MHz.
- It comes with a strong magnetic mount. It also has good coax.
- It has a rubber washer that fits over the space between the antenna's base and plug.

2 Pack-29 Inches Foldable/Tactical Radio Antenna

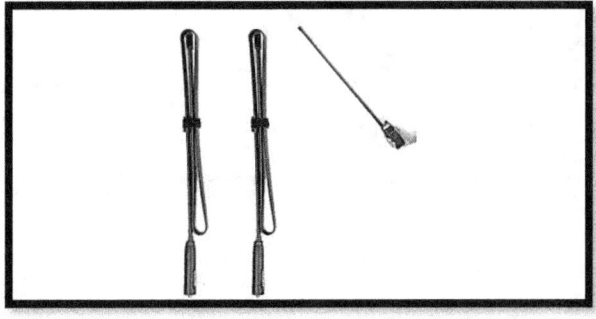

For the price, it's a good antenna. If what you want is a small antenna, this is a good choice. Also, the antenna is bendable, so you don't have to worry about the link getting caught or broken. It's great that it can be folded up. In a military setting, this antenna works well. This easy piece of gear could be a great choice for you if you want to go on a journey or do something outside. With good range and clarity, it's one of the best antennas for the Baofeng UV-5r. It's much easier to set up this antenna than a stock whip antenna. The build quality and materials used will make you feel like you're in the military. As before, it is a radio device that will improve your ability to talk to people far away. The style doesn't look good, and it could have been a little better. It's just a small issue, though, and it doesn't affect the speed at all.

Key Features

- It's easy to fold up and can be used outside.
- It works with frequencies between 136 and 174MHz and 400 and 520MHz
- It's perfect for the Baofeng UV 5R for a huge range boost when fully extended;
- SMA-Female connector

3. Dual Band 136-174 MHz&400-520 MHz Antenna for Baofeng UV-82 UV-5R

Two-band 136-174MHz and 400-520MHz SMA-female antenna for Baofeng UV-82 UV-5R is the next item on the list. It's something that makes the Baofeng radio better at both receiving and sending. It has a lot more gain than an OEM antenna. The antenna is made up of two parts that are linked with a SMA female connector. From what I can tell, it's well-built and strong, so it should be able to do its job anywhere. Because it's so light, it's just an antenna that you can take with you. Not only that, but these two antennas are great for the price. There is a chance that you will have trouble connecting these radio pieces. It won't make a big difference, though.

- It works with some Baofeng mobile ham radios, such as the GT-3, UV-5R, UV5RE, UV5RA, UV5R Plus, UV-82L, and BF-A58.
- Its frequency range is 136–174 MHz and 400–520 MHz.
- Package includes SMA-female link

4. 42.5-Inch Length ABBREE SMA-Female Dual Band Foldable CS Tactical Antenna

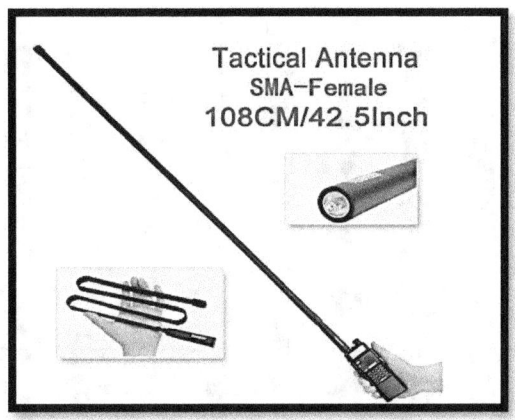

The next antenna is a 42.5-inch long ABBREE SMA-Female dual band 144/430 MHz foldable CS antenna. When you fold this antenna up, you can carry it with you in your purse. It works well with the Baofeng BF-F8HP and is one of the best antennas for it. The antenna is made of two parts. A screw holds the base to the radio, and another screw holds the antenna to the base. Would you like to take it outside to work in the field? Cool!!!! In the field, it works better and has a stronger signal thanks to this great antenna. You need an antenna that works well on both UHF and VHF, right? Ok. We get it. On both UHF and VHF, it works just as well. What a great thing it is!!! Doesn't it? Do you like to hike? If so, it's a great deal for you. You know a large bad antenna is much better than a well-tuned small unit. As a result, it works better and has a wider range than the stock unit or most alternative units. But the fact that it looks silly bothers me a lot. The SMA female connection is also not very strong. But taking into account its price and benefits, it might be a good deal for you.

Key Features

- It has a Velcro strap and rubber washers.
- It works well with Baofeng radios like the UV-5R, UV-82, and BF-F8HP Ham.
- The frequency range is 144 to 430 MHz.
- An antenna that is soft and less than 1.5 V.S.W.R.

5. BaoFeng bf-888s 10 X Original Antenna for BaoFeng Bf-888s

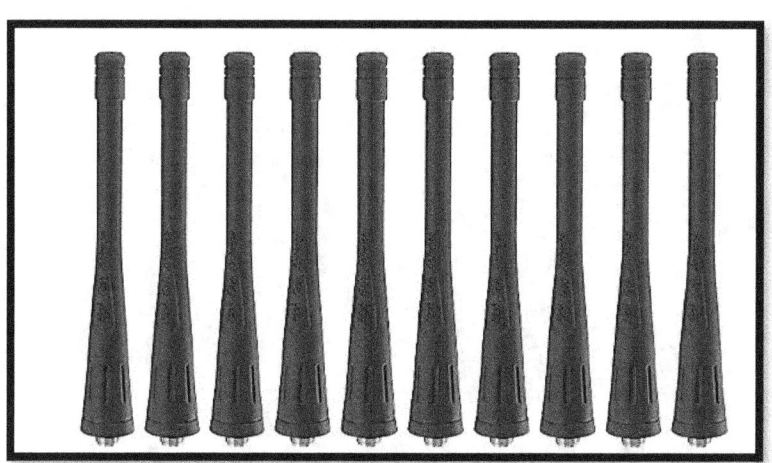

The BaoFeng bf-888s Antenna is another great antenna for Baofeng that we've included on this list. It is one of the best antennas for Baofeng BF-888S that you can get. If you need more than one use for various radios, this antenna meets your needs by offering various uses for various radios. For the price, this radio is a great buy. But it's not long enough.

Key Features

- The frequency range is 400–470MHz, which can receive and send signals perfectly in the field.
- Works with different radios, such as BF-888s and uv-5r
- Strong design made of strong materials

6. Bingfu Dual Band VHF UHF Ham Radio Antenna

The Bingfu Dual Band VHF UHF 136-174MHz 400-470MHz Ham Radio Antenna is the next item. Bingfu has been making antennas for a long time and is very good at it, especially when it comes to radio communications. This compact antenna from Bingfu works better than any other antenna on the market. The fixing surface is made of stronger steel, which shows that it is well-built.

Want a good range? This antenna's range is longer and it covers more area. It also looks cool. Want to carry an antenna around with you? So, it's a good deal because it's small and easy to carry, even on planes. Once more, it's a great antenna for the Baofeng HT mobile app. This is one of the best antennas for the Baofeng UV-82 that you can find. The base is about 1 inch wide. The cord is too short, though. Also, the magnet mount isn't very strong.

Key Features

- Strong signal between 136 and 174 MHz on VHF mode and 400 to 470 MHz on UHF mode
- It has a piece of double-sided tape that you can use to connect the antenna to your radio.
- It can handle up to 75 watts
- It works with BaoFeng radios BF-F8HP, UV-5R, UV-82, and BF-888S.

7. Baofeng Magnetic Car Vehicle Mounted Antenna

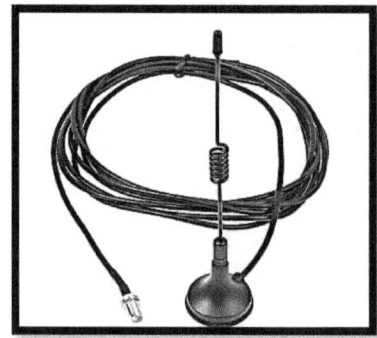

The next item on the list is the Baofeng Magnetic Car Vehicle Mounted Antenna UHF VHF Dual Band. The Chinese company Baofeng is one of the biggest and most popular in the world, and they make great amateur radio gear for their users. This is one of the best long-range antennas for the Baofeng UV-5R that you can find. The quality of the build looks good. It has two bands, but it mostly works with UHF. It's too bad that the base magnet is so weak.

Key Features

- ❖ It has a gain of 2.15dBi and a frequency range of 400 to 470MHz.
- ❖ Because it has a coaxial connection, it can connect radios like the BAOFENG UV5R Plus UV5RA Plus UV3R Plus
- ❖ With less than 1.5 VSWR, the most power that can be sent out is 5 watts.

8. BaoFeng SRH805S SMA-F Dual Band Antenna

The Baofeng SRH805S SMA-F SRH805S SMA-F Female Dual Band Antenna is another great antenna made by Baofeng. We've already said that Baofeng is a great brand that makes high-quality radio electronics. The brand is from China. The Original SRH805s is small and easy to carry, making it good for work in trees. The size of it is bigger than the stock duck. A lot of people do say, though, that the antenna is too small. It would have been better if it was longer.

Key Features

- ❖ The frequency range is 136-174 MHz on VHF and 400-520 MHz on UHF
- ❖ Fits for Baofeng GT-3, UV-5R, BF-888s Radio
- ❖ SMA-F Female connector

9. Walkie Talkie Antenna 15.6-inch Whip Dual Band UV

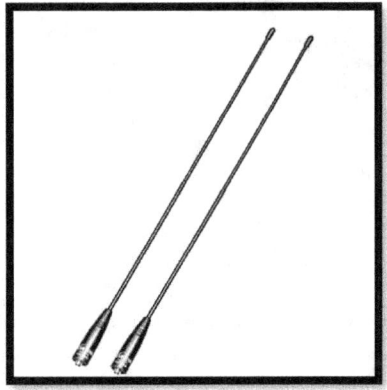

The Walkie Talkie Antenna 15.6-inch Whip Dual Band UV from LUITON is the next best Baofeng antenna on our list. The Nagoya NA-771 and this antenna work pretty much the same way. The good thing about it is the clear sound. The plastic washer at the base of this antenna is one of my favorite parts. It lets you seal the antenna and HT together. To go over this one more time, the antenna seems to have great mechanics—better than most stock antennas. The antenna is protected by a plastic shell, and the base is made of metal. This makes it very good for use in the field. Would you like an antenna that is easier to set up and install? Okay!!!! Setting up and putting this antenna in is very easy. You can find a way to enjoy the great broadcast and receiving. It is, however, a stronger receiver.

Key Features

- ❖ The frequency range is 144–430MHz, which lets you receive and send signals over a large area.
- ❖ VSWR is less than 1.5:1 and the most power that can be sent is 10 watts.
- ❖ It works with UV-82, UV-B5, GT-3, BF-F8HP, UV-5RA, UV-5RE, and UV-5R.
- ❖ It comes with two rubber rings.

10. HYS SMA-Female Handheld Dual Band Antenna

This is the last best antenna for Baofeng on our list. It is the HYS SMA-Female Handheld Dual Band Antenna. For a long time, HYS has been making great radiocommunications goods. It has two parts, and you can fold them up. That makes it easy to bring in your purse on your trip. It doesn't matter if you are going camping or shooting. This antenna will work no matter what you face on your trip because it is reliable and dependably strong. Plus, it's one of the best antennas for baofeng uv 82hp out there. However, this antenna doesn't work well enough in VHF mode.

Key Features

- ❖ It has a frequency range of 144 to 430 MHz.
- ❖ It can take up to 20W of power.
- ❖ It works with BaoFeng uv-5r, uv-82, f8hp, and BF-F8+ radios.

Things to Consider Before Buying the Best Antenna for Baofeng

When the market is full of the same items that are good in different ways, it can be hard to find the best one. If you want to know what the best Baofeng antenna is, you should do a lot of math in your head. So, a good antenna is needed to improve the network for Baofeng Ham Radio. Read these things to find out what makes a Baofeng antenna the best.

Efficiency

People always want to buy a Baofeng antenna because of how well it works. How useful is it to have an antenna? Antenna theory says that antenna efficiency means transmission efficiency. Antenna efficiency in radios is the amount of electricity that the antennas use to turn radio frequency power into emitted power. So, explaining the antenna's effectiveness is a matter that can ease all your worries about the Baofeng antenna's performance.

Antenna Compatibility

An antenna should work with all Baofeng ham radio types. When you lose your radio. You need a new radio, but the type of the one you had before is no longer available. This is why you move to a different type. Are you also going to change the antenna? Without a doubt, it will cost more. A lot of people have this case, and they're wasting money on it. What's the point of spending more? Be smart. If you buy a Baofeng radio antenna that works with all Baofeng models, it will work with all Baofeng models that come out after that.

Durability

If everything works well with your next Baofeng antenna, you should think about how long it will last. How can I find out if it will last? You know how easy it is to learn. For an antenna to last, it must be well-made and made of the right materials. So first, make sure that the materials used to make your next antenna are strong and that the design is also strong.

Size

Another thing that should be carefully looked at is the antenna's size. Size is very important, especially when sending and receiving from or to a faraway place. A longer antenna lets you receive signals better from far away and send signals more clearly to faraway areas. It is portable, though, which many people like. The longer antenna is not a good idea in this case. In that case, the shorter antenna is a clear winner.

CHAPTER 13

UNDERSTANDING ANTENNA GAIN AND POLARIZATION

Antenna gain is the amount by which broadcast and received signals are amplified or concentrated. This makes the signals stronger and wider. Knowing how antenna gain works can make it a lot easier for amateur radio operators to make solid links over long distances. Think about a made-up situation in which a radio operator in a distant mountain area wants to talk to other radio fans hundreds of miles away. With limited tools and rough terrain, it's important to make sure that circuits are clear and don't break. This is where antenna gain comes in, giving us possible ways to get around problems like weak signals caused by distance or natural hurdles.

Understanding Antenna Gain

Imagine that you are getting ready for a long-distance call. You have set up your gear and tuned it to the frequency you want to use, but antenna gain is still the most important thing that will decide how well your transmission goes. This part will talk about the idea of antenna gain and how important it is for getting the best signal receiving. When an antenna can focus sent or received information in a certain direction, this is called antenna gain. It measures how much power an antenna sends out in a certain direction compared to an ideal isotropic radiator, which sends out the same amount of power in all directions. To better understand this, let's look at a real-life example: Picture two antennas next to each other that send signals at the same amount of power. What would you be able to tell the difference between them if they had different gain levels? The antenna with higher gain would send more energy in the desired direction, making the signal stronger and covering more ground over longer distances.

To understand why antenna gain is so important, you need to know what it does for you:

- ❖ **Better Signal Strength**: Antennas with higher gain can boost signals from weaker sources, making it easier to talk to each other.
- ❖ **Enhanced Range**: Antennas with higher gain can reach farther away than those with lower gain because of their increased directed focus.

- ❖ **Less Interference**: High-gain antennas reduce interference from nearby sources or annoying noise by narrowing their transmission pattern.
- ❖ **Better Reception Quality**: A better signal makes it possible for conversation meetings to have clearer sound with less confusion.

The following table shows the different types of common antennas and their related gains:

Type	Gain (dBi)	Applications
Yagi-Uda	10	Long-range point-to-point links
Dipole	2	General-purpose, short-range usage
Parabolic	30	Satellite communication
Log-Periodic	8	Wideband communications

Understanding antenna gain is very important because it has a direct effect on how well their communication system works. Operators can get the best signal strength and range with the least amount of clutter by choosing the right antenna with the right gain.

Types of Antenna Gain

Understanding Antenna Gain is important for people who want to improve their ability to communicate. We talked about the idea of antenna gain and how it can help improve signal transfer and receiving in the last part.

Now, let's talk more about the different kinds of antenna gain that are usually used.

- ❖ **Directivity**: Increasing the antenna's directivity is one way to boost its gain. In this case, the energy pattern is focused in a certain direction, which makes the signal stronger in that area while weaker elsewhere. For example, let's say you are an amateur radio operator on top of a mountain and you are trying to get in touch with a station that is far away but at sea level. You can focus your broadcast energy on a specific place by using an antenna with high directivity gain that is very directional. This greatly increases your chances of communicating clearly.
- ❖ **Efficiency**: Another important thing to think about when thinking about antenna gain is how well an antenna turns input power into electromagnetic waves that travel through the air. More effectiveness means that the information is stronger

and can reach farther. To show this, let's say that two antennas have the same gain (dB), but one is more efficient because it was better designed or made of better materials. It will then send a greater signal than the other.

❖ **Beamwidth**: The beamwidth tells you the angle at which an antenna can send or receive messages without losing a lot of performance. A smaller beamwidth means more directivity, which means more gain in that small covering area. On the other hand, a bigger beam width would give up some directivity in exchange for wider coverage, but it would also lose some gain per unit area covered.

❖ **Front-to-Back Ratio**: The front-to-back ratio shows how well an antenna can tell the difference between signals coming from the right way and those coming from behind it. It gives a number that tells you how much more sensitive an antenna is to messages coming from the way it wants to receive them than from other directions. A high front-to-back ratio means that useless signals are better thrown away, which leads to better gain and signal clarity.

▪ Increased directivity allows for more focused transmission and reception.

▪ Higher efficiency leads to stronger signals and greater range.

▪ Narrower beamwidth provides higher gain within a limited coverage zone.

▪ A high front-to-back ratio ensures better discrimination between desired and undesired signals.

Markdown Table

Type	Explanation
Directivity	Focusing the radiation pattern towards a specific direction
Efficiency	The ability of an antenna to convert input power into radiated electromagnetic waves
Beamwidth	The angle over which an antenna can effectively transmit or receive signals without significant loss in performance
Front-to-Back Ratio	Measures the ability of an antenna to discriminate between signals received from the desired direction versus those coming from behind it

In summary, radio fans who want to improve their speaking skills must understand the different types of antenna gain. Operators can choose antennas that give them stronger signals, longer range, and better total performance by looking at things like directivity, efficiency, beamwidth, and front-to-back ratio. After looking at the different parts of antenna gain, let's move on to the things that affect it to get a better grasp of the subject.

Factors Affecting Antenna Gain

Now, let's talk more about the different kinds of antenna gain you should know about. One type that stands out is **directive gain**, which means that an antenna can move energy in a certain way. This can make a big difference in the power and quality of the signal for long-distance contact. For example, let's say you're trying to get in touch with another ham user who lives on a faraway island. If you use an antenna with a high directed gain, you can focus the energy you send towards that specific spot, which increases the chances of communicating successfully.

Reflector gain is another type. This type uses extra parts or structures to improve the general radiation pattern and make the signal stronger. The radiated energy is sent in the way that is wanted by these mirrors, which also cuts down on radiation that isn't wanted in other areas. It's like carefully putting mirrors around your antennas to boost messages where they mean the most. We also have **parasitic element gain**, which is also called "the power of companionship." This type takes advantage of how multiple antennas work together by adding extra parts called directors or mirrors close to the main driven element. Because they are carefully placed and tuned, these extra parts help focus and direct the energy that is emitted better than a single antenna could do by itself.

To summarize:

- ❖ **Directive Gain**: Focuses energy in a certain way to improve contact over long distances.
- ❖ **Reflector Gain**: This type of gain uses extra parts or structures to make radiation patterns better.
- ❖ **Parasitic Element Gain**: This feature improves performance by letting close antennas interact with each other.

It is helpful for amateur radio users to know about these different types of antenna gain so they can choose the best tools for their needs. Next, we'll talk about how to measure antenna gain, which is an important part of figuring out how good different antennas are at what they do.

Measuring Antenna Gain

The antenna's size is one of the most important things that affect its gain. Most of the time, bigger antennas have higher gains because they can pick up more electric energy from their surroundings. However, it's important to keep in mind that sometimes it might not be possible or practical to make an antenna bigger. The design of the antenna is another important factor. Gain amounts change between different forms. For example, a Yagi-Uda antenna usually has high gain and directivity for point-to-point contact, which means it can be used for long-distance sending. Dipole antennas, on the other hand, send signals in all directions, but they might not have as much gain as directed antennas. The frequency at which an antenna works is also a big part of how much gain it has. Antennas are usually made to work with certain frequency ranges, and how well they work changes as the frequency range changes. To get the most out of an antenna's gain potential, it's important to pick one that works with the frequencies you want to use. Now, let's make up a story about John and Sarah, two radio fans. For their sites, they both decided to set up HF (high-frequency) antennas.

Here is a table that compares some of the things that affect antenna gain:

Factor	John's Antenna	Sarah's Antenna
Physical Size	Large	Small
Design	Dipole	Yagi-Uda
Operating Frequency	3-30 MHz	14-30 MHz
Expected Gain Potential	Moderate	High

In this case, John chooses a bigger dipole antenna that works with a wider range of frequencies, while Sarah picks a smaller Yagi-Uda antenna that has a higher gain potential at the frequencies she wants to use. If you know what affects antenna gain, you can choose the right antenna for your needs. Picking the Right Antenna Gain is a key part of getting the most out of your wireless connections. Now, let's talk about some important things to think about and some tips that will help you make a smart choice.

How to Pick the Best Antenna Gain

Let's say you're excited to test out your new antenna for your amateur radio station because you just finished building it. One way to figure out how good an antenna is is to measure its gain. This part will talk about the different ways that radio fans can correctly measure antenna gain.

So, let's start by looking at some popular ways to measure antenna gain:

❖ **Measurement of Field Strength**: A standardized field strength meter can be used to find out how strong the signal is at different distances from the antenna. It is possible to figure out the tested antenna's gain by comparing these numbers to those from a reference antenna.

❖ **Single-Antenna Method:** Here, there is only one antenna and a reference source whose power output and radiation pattern are known. Different directions around the tested antenna are used to measure the received power, which lets the exact gain of the antenna be calculated.

❖ **Two-Antenna Comparison**: Using two antennas, one as a reference and the other as a test and carefully limiting things like distance and direction, it is possible to make direct comparisons between them using tools made just for that purpose.

Now that you know about some common ways to measure, let's look at how picking the right antenna gain is a key part of getting the best results.

Embrace Better Performance

– Unlock increased range

– Enhance signal clarity

– Improve overall reliability

– Maximize communication capabilities

To get the best results from the antenna gain we choose, we need to think about a few key things:

	Gain (dBi)	Range (Miles)	Signal Clarity	Reliability (%)
Antenna A	10	50	Satisfactory	90
Antenna B	15	75	Highly clear	95
Antenna C	20	100	Crystal clear	98
Antenna D	25	125	Exceptional	99.5

By looking at the table above, it's clear that antenna gain has a direct effect on range, signal quality, and general dependability. In our search for better performance, picking an antenna with a higher gain can let us communicate with more people. Now that we have a good grasp on how to correctly measure antenna gain and how choosing the right gain number can affect how well a system works, we can move on to finding even better ways to optimize antenna gain.

Optimizing Antenna Gain for Better Performance

Continuing smoothly from the last part on picking the right antenna gain, let's now talk about how to make the antenna gain work better. Let's say you put up a high-gain Yagi antenna to help your amateur radio station receive signals better. Despite its great specs, however, you are not seeing the expected improvement in signal strength. To solve this problem and get the most out of your gear, you need to find different ways to improve radio gain. First, you might want to change your antenna's height and direction. By raising an outdoor antenna higher above the ground, you can get rid of things like nearby buildings or plants that might get in the way of signal transmission. Trying out different azimuth angles (turning the antenna so that it faces the opposite direction) can also have a big effect on signal strength by finding the best way to get the signal to the emitter. Second, using the right feedline lengths and plugs can help your receiver or transmitter and antenna lose less information. Power can be sent over long distances without losing much of its strength if the wires are of good quality and have low loss. Using well-matched plugs on both ends of the wire also stops impedance mismatches that could hurt the performance of the whole system.

Finally, adding active devices like preamplifiers or antenna amplifiers close to your receiver can boost weak signals even more before they get to it. These devices improve received signals while reducing any noise that might be introduced during transfer over long wire runs. When choosing the right amplification settings, care should be taken so that too much noise or distortion doesn't get into the received data.

To sum up these methods for optimization:

- ❖ Changing the antenna's height and direction
- ❖ Using the right lead lengths and connections for the feedline
- ❖ Active devices for signal intensification are being added.

By using these tips correctly, you can get the most out of your chosen high-gain antenna setup and get better overall performance in the form of longer range, clearer signals, and less interference.

Optimization Technique	Benefits	Considerations
Adjusting antenna height and orientation	Minimizes obstructions, improves range	Requires suitable mounting
Using proper feedline lengths and connectors	Reduces signal loss over distance	Select low-loss coaxial cables with matched connectors
Incorporating active devices during transmission	Amplifies weak signals, reduces noise selection	Careful amplification level

By using these techniques to get the most out of your antenna gain, you can make your amateur radio experience a lot better. If you can improve your coverage, you will be able to speak more clearly and listen to a wider range of bands, which will make your time on amateur radio more fun.

Understanding Antenna Polarization

Antenna polarization is one of the most important ideas in amateur radio because it has a big effect on how signals travel and how they are received. Understanding how radio polarization works can make it a lot easier for fans to talk to each other over long distances. For example, let's look at an amateur radio user who lives in a city with lots of people and a lot of electromagnetic pollution. This user could improve their chances of getting a clear signal even when there was a lot of noise around by carefully choosing the right antenna orientation. In the world of amateur radio, antenna orientation is very important. It has to do with the direction of the electromagnetic waves that an antenna sends or receives. Understanding antenna orientation is important for getting the best signal transfer and receiving because it has a direct effect on how well communication works. To show this idea, let's make up a story: two amateur radio operators are talking to each other through mobile transceivers. Operator A holds the receiver vertically, with the antenna straight out from the body. Operator B holds theirs laterally, with the antenna straight out from the body. Because they are facing different ways, the signal strength and clarity may be different for each.

- ❖ The orientation of an antenna affects how well messages can be sent and received.
- ❖ Some of the most popular types of radio orientation are circle, horizontal, and vertical.
- ❖ Having the polarizations of the sending and receiving antennas match makes the system work more efficiently.
- ❖ Antenna orientation can be changed by interference from close items or changes in the weather.

A table can also be used to show details about antenna polarization:

Antenna Polarization Types	Characteristics
Vertical	Suitable for long-distance communications
Horizontal	Effective for short-range communications
Circular	Provides both vertical and horizontal coverage

Radio fans can make wise choices about how to set up their tools by understanding the significance of antenna orientation. By arranging the sending and receiving antennas correctly, the information can travel as far as it can go. Transitioning into why antenna polarization is important for radio enthusiasts brings us closer to exploring its practical implications without explicitly stating *"step"*.

Why is antenna polarization important for amateur radio?

Let's think about a made-up situation to get a better idea of what antenna orientation means in real life. Let's say that Alex and Ben, two amateur radio operators, are talking to each other using small VHF radios from 100 miles away. Both users are sending on the same channel and have antennas that are the same. In this case, Alex and Ben both set their antennas to vertical polarization at the start. But they quickly learn that there is a lot of clutter and signal loss in their contact. They are so fed up that they decide to try changing the antenna's orientation. Alex changes the orientation of his antenna to horizontal, while Ben keeps it vertical. Suddenly, they can hear each other perfectly, with no noise or interruption to be seen. This simple change made a huge difference in how well their transmission worked. This case study shows how important antenna orientation is for people who like to listen to the radio.

Here are some important facts about antenna polarization that will help you understand how important it is:

- ❖ Antenna polarization is how the electric field of an antenna is oriented to the ground.
- ❖ The polarity of an antenna can have a big effect on the strength and quality of a signal.
- ❖ Matching the polarizations of the send and receive antennas improves the speed of data movement.
- ❖ There are four different kinds of antenna polarizations: vertical, horizontal, circular (right-hand), and circular (left-hand).

Type	Advantages	Disadvantages
Vertical	Omnidirectional pattern	Susceptible to noise
Horizontal	Reduced noise	Limited range
Circular RH	Better multipath	Lower gain
Circular LH	Improved satellite comms	Less common equipment

Radio enthusiasts can choose and change the polarizations of their antennas with confidence if they understand these ideas and think about how they can be used in real life, like in our case study example. Moving on to the next part, "**Different types of antenna polarization**," we will talk more about what makes each type unique and how it can be used. Understanding these differences is important for getting the best results in a variety of radio contact situations.

Different types of antenna polarization

Now, let's look at the different types of antenna orientation that are often used in home radio stations.

By knowing about these differences, you can pick the option that best fits your wants.

1. Vertical Polarization:
   - In this configuration, the radiating elements of the antennas are oriented vertically.
   - When talking to ground-based stations or mobile stations on roads and highways, vertical polarization is often used.
   - This kind of polarization works well when low-angle transmission is needed, like when talking on the phone over long distances over flat ground.

2. Horizontal Polarization:
   - When antennas have horizontal polarization, their emitting parts are laid out flat on the ground.
   - This position is often used for contact via line of sight between fixed stations that are about the same height above the ground.
   - It works especially well in cities, where vertically polarized antennas can lose signal strength because of buildings and other obstacles.

3. Circular Polarization:
   - In circular polarization, the electric field vector of an electromagnetic wave is rotated around its axis while keeping its size the same.
   - It's helpful because it reduces problems caused by multipath interference and improves signal reception from multiple sources at the same time.
   - Satellite communications, mobile systems, and environments with a lot of signal echoes all use circularly polarized antennas.

Depending on things like distance, geography, and hurdles in the way, using different types of antenna orientation has different benefits. If you carefully think about these choices, you can make your amateur radio setup work better without changing anything. Moving forward, let's delve into a discussion about factors that impact antenna polarization effectiveness.

Factors affecting antenna polarization

After talking about the different types of antenna polarization that are popular in amateur radio sets in the last part, let's now talk about the things that affect the choice of antenna polarization. Having this knowledge is very important because it lets fans communicate more effectively based on the situation. Consider a made-up situation in which an amateur radio operator wants to talk to another operator hundreds of miles away over long distances. Because it's sensitive to sun flares and other natural events, the frequency band that was picked is known to have high amounts of ambient noise. In this case, picking the right antenna orientation is very important to keep signal loss to a minimum and receiving quality at its best.

Four main types of things can change the orientation of an antenna:

1. Environmental Conditions:
   - Presence of nearby buildings or structures
   - Terrain characteristics (e.g., mountains, valleys)
   - Vegetation density (e.g., forests, urban areas)
2. Propagation Characteristics:
   - Ionospheric conditions (affected by time of day and sunspot activity)
   - Atmospheric effects (e.g., refraction, scattering)
3. Frequency Band Considerations:
   - Wavelength relative to object size
   - Signal penetration through obstacles
   - Absorption properties of different materials
4. System Requirements:
   - Desired coverage area
   - Required signal strength at receiver location(s)
   - Signal interference considerations

Please see the table below for more information on how each of these factors affects the choice of antenna polarization:

Category	Influence
Environmental	Nearby structures may cause multipath reflections leading to fading; vegetation affects signal loss
Propagation	Varying ionospheric conditions affect signals differently; scattering impacts overall signal quality
Frequency Band	Higher frequencies tend to require vertical polarization for efficient ground wave propagation
System Requirements	Directionality requirements affect choice; interference concerns may necessitate specific options

Radio fans can choose antenna orientation with confidence if they are aware of these factors. Operators can improve the range and dependability of their communications by carefully thinking about how their setup is set up.

How to determine the polarization of an antenna

We talked about how antenna polarization is important in amateur radio and how it can change how signals travel in the last part. Let's look more closely at the different things that affect antenna orientation. The working cadence is an important thing to think about. For best performance, different bands need different types of antennas with specific polarization directions. For instance, VHF (Very High Frequency) and UHF (Ultra High Frequency) frequencies usually use vertically polarized antennas because they are commonly used in mobile communications and land-based rebroadcast systems. Another important factor is how close the item or building is to it. Reflections and multipath interference can be caused by nearby buildings, trees, or even uneven ground, which can change the polarization features of an antenna's broadcast pattern. When placing your antenna to get the most out of it, it's important to know about these surroundings.

Changing the communication medium will also cause a change in the polarization direction because that's how electromagnetic waves behave. This idea is useful when moving from space to things like wires and plugs. Keeping the polarization the same across the whole system is important to keep signal loss to a minimum and make sure transmission works well.

To better show how these things affect antenna polarization, let us look at an example:

Let's say you are putting up an amateur radio station in an area with a lot of big buildings. This time:

❖ When tall buildings are close together, they may bounce messages with different polarizations.

❖ It is more likely that echoes would cause multipath interference in vertical antennas.

❖ If changes need to be made, the vertical antennas could be tilted a little, or horizontally polarized antennas could be considered instead. By doing good site studies before installation, problems that might affect polarity could be found.

❖ The following table lists some important things to think about when choosing the right antenna orientation for different situations:

Scenario	Recommended Antenna Polarization
Urban areas with tall buildings	Horizontal
Open rural environments	Vertical
Mobile or portable operations	Circular (Omnidirectional)
Satellite communications or weak signals	Crossed (Circular/Linear)

In conclusion, radio fans need to know about the things that affect antenna orientation. One can improve signal travel by taking into account working frequencies, the environment, and the qualities of the communication medium.

Practical applications of antenna polarization in amateur radio

Moving on from the last part, which talked about how to figure out an antenna's polarization, let's look at how antenna polarization is used in amateur radio. For example, let's pretend that two operators are trying to talk to each other over a long distance using small VHF radios. This will show how important it is. Operator A has an antenna that is polarized vertically, and Operator B has one that is polarized horizontally.

For communication to work in these situations, you need to know how antenna polarization changes signal propagation. Here are some important things to think about:

- **Signal strength**: The orientation of the antenna has a direct effect on signal strength by increasing or decreasing signal receiving. When the sending and receiving antennas have the same polarization (vertical-to-vertical or horizontal-to-horizontal), the signals that are received are much stronger than when the polarizations are not matched (vertical-to-horizontal or vice versa).
- **Less interference**: Using antennas with different polarizations can cut down on interference from close sources sending out unwanted messages. This can improve the quality of conversation generally and cut down on interruptions during events or games with a lot of people.
- **Limiting multipath fading**: Multipath fading happens when signals bounce off of buildings, terrain features, or the atmosphere, causing multiple signal lines to reach the listener at the same time with different phases. Using antennas with different polarizations can help solve this problem, since signals coming in from different directions may fade at different rates depending on their polarities.

Regarding diversity reception, using antennas with different polarizations in diversity reception systems makes them work better in tough situations, like cities or places that get a lot of bad weather. These systems move between several antennas to pick the one that sends the best and least influenced signal at any given time. This makes the system more reliable and increases its range.

To make these real effects even clearer, look at Table 1, which compares different factors that affect matching and mismatching antenna polarizations?

Factor	Matching Polarizations	Mismatched Polarizations
Signal strength	Stronger	Weaker
Interference reduction	Enhanced	Reduced
Multipath fading	Minimized	Increased
Diversity reception	Improved performance	Limited effectiveness

In conclusion, radio enthusiasts who want to communicate reliably and effectively need to know how antenna orientation can be used in real life. Operators can get better results when setting up links over long distances or in difficult situations by thinking about signal strength, interference reduction, multipath fading prevention, and diversity receiving.

CHAPTER 14
CHOOSING THE RIGHT PROTECTIVE GEAR

When choosing defensive gear for using Baofeng radios, there are a few things to keep in mind to ensure safety, ease of use, and the best performance. Protective gear can include a wide range of extras and tools that make the radio easier to use and safer while it's being used.

Some ideas are given below:

1. Radio Case or Holster:
   - ❖ **Carrying Case or Holster**: This will keep the radio safe from bumps, scratches, and dust while making it easy to get to. Look for bags or holsters that are made to last and are made to fit your Baofeng radio type.

2. **Antenna Protection**:
   - ❖ **Antenna Protector**: Think about getting a cover or cap for the radio's antenna to keep it from getting damaged or bent when it's not in use or while it's being moved.

3. **Screen Guard:**
   - ❖ Put a screen cover on the screen to keep it from getting scratched or broken, especially if you use the radio in rough places.

4. **Battery and Charger Safety:**
   - ❖ **Extra Batteries**: Bring extra batteries with you in case you need to use them for a long time, like when you're camping or in an emergency.
   - ❖ **Charger and Battery Care**: To make sure charging is safe and effective, use a charger that works with your batteries and handles them the right way. Do not overcharge batteries or use cells that are broken.

5. **Hands-Free Operation:**
   - ❖ **Headset or Earpiece**: If you want to talk on the phone without using your hands, get a headset or earpiece that has a microphone built-in. This lets you talk without being seen and gives you free hands for other things.

6. Environmental Protection:

❖ **Water-Resistant Bag or Cover**: If you use the radio outside or in bad weather, you might want to think about getting a waterproof bag or cover to keep water and dirt off of it.

7. **Extended Antenna Options**:
   ❖ **External Antenna**: If you want better range and coverage, look into external antenna choices that work with your Baofeng radio model. Make sure it works with other things and follows the rules in your area.

8. **Manuals and Guides**:
   ❖ **User Manual and Guide**: Keep the user manual close by for reference, to fix problems, and to learn about the radio's features and limits.

9. **Personal Safety Gear**:
   ❖ **Weather-Appropriate Gear**: If you're using the radio outside or in a tough setting, you might want to bring extra personal protection gear like weather-resistant clothes, gloves, or a helmet.

10. **Following the rules**:
   ❖ **License and rule compliance**: Make sure that the way you use the radio doesn't break any local rules or license requirements to stay out of trouble with the law.

When picking out safety gear for Baofeng radios, you should think about the model, how you plan to use it, and the weather to make sure that the items you choose make the device easier to use, keep it safe, and help you communicate effectively and safely.

Accessibility and Comfort Considerations

Accessibility and comfort for Baofeng radios should be thought about in some ways to make sure they are useful and comfortable for a wide range of people.

Here are some things to think about:

1. **Physical Accessibility**:
   • **Button Size and Placement**: Make sure that the buttons are spaced out evenly and are the right size for easy use, especially for people who have trouble with precision or fine motor skills.
   • **Tactile marks**: People who are blind or have low vision can use important buttons with tactile markers or higher marks to help them.

- **Easy-to-Grip Design**: The radio should be easy to hold and move around, with a material or shape that makes it comfortable to hold.

2. **User Interface and Controls**:
   - **Simple Interface**: To keep things simple, make sure the interface is easy to understand by using clear labels and buttons.
   - **Voice Prompts**: Voice prompts or audio feedback can help people who are blind or have low vision find their way around settings and features.
   - **Adjustable Settings**: Give people with different hearing skills the chance to change the volume, tone, and other settings.

3. **Display and Feedback**:
   - **High Contrast Display**: For people who have trouble seeing, a display with high contrast and lighting that can be changed can be helpful.
   - **Auditory Feedback**: Users who are blind or have low vision can benefit from hearing alerts or beeps for important tasks like turning on/off or changing stations.

4. **Comfort and Wearability**:
   - **Comfortable Ways to Carry**: For better transport, think about adding straps or clips as extras.
   - **Weight and Size:** Make sure the radio is light and the right size so you can use it comfortably for long amounts of time.

5. **Battery Life and Charging**:
   - **Large Battery Life**: A larger battery life can be helpful for people who have trouble charging their devices often.
   - **Accessible Charging Ports**: Make sure that the charging ports are easy to get to and work with different charging ways, like USB charging.

6. **Language and localization**:
   - **Support for multiple languages**: If you can, give people from a wide range of linguistic groups the chance to choose a language.
   - **Localization Settings**: For user ease, think about regional settings like date/time forms.

7. **Help and documentation**:
   - **Easy-to-Follow Directions**: Give users complete, easy-to-find books or guides with clear pictures and instructions.
   - **Customer Support Accessibility**: Make sure that customer service is open and easy to reach for users who may need help or have specific questions about accessibility.

8. Compliance with Accessibility Standards:

- **Compliance with Accessibility Guidelines**: Make an effort to follow accessibility rules and standards to make sure that all users can use the site.

Taking these things into account can help make Baofeng radios easier for more people to use and more comfortable for all of them, making them more user-friendly and welcoming.

Hardware Modifications

Baofeng radios can have their hardware changed to make them easier to use, more accessible, or more useful. But it's important to keep in mind that changing electronic devices could void guarantees and, if not done properly, could damage the device.

People might want to make the following changes to their hardware:

- **Extended antennas**: Getting a longer or higher-gain antenna can make it easier to receive and send signals over longer distances. This can be especially helpful for people who need a better connection.
- **External Speakers or mics**: Adding speakers or mics from the outside can improve the sound quality and make it easier for people to talk to each other in noisy places.
- **Battery Upgrades**: Getting portable or higher-capacity batteries can make the radio last longer, which may be useful for some users.
- **Belt clips or holsters**: Adding or changing belt clips or holsters can make bringing the radio easier for people who need to get to it quickly while they're on the go.
- **Braille Labels or Tactile Lines**: Putting Braille labels or tactile lines on buttons and settings can help people who are blind or have low vision use the gadget.
- **Custom Cases or Grips**: You can make or buy custom cases or grips that make the radio easier to hold or carry and improve its usability.
- **Charging Accessories**: Buying charging stands or other charging options can make it easier to charge the device.
- **Frequency Expansion or Changes**: Some users may try changes to make frequency ranges bigger or to get access to features that aren't offered by default. But it's important to know what the laws are in your area and if you need a license to use certain radio bands.

- **Ruggedization or waterproofing**: Putting protected cases or seals on the radio to make it less likely to get damaged by water, dust, or drops can make it last longer in some settings.

Before making any changes, it's important to do a lot of study, follow the manufacturer's instructions (if they're available), and, if necessary, talk to professionals or other knowledgeable people who work with radio electronics. Also, always make sure that any changes you make to radio devices are in line with the laws, rules, and safety standards that apply to their operation and use.

CHAPTER 15
INTERNATIONAL RADIO REGULATIONS
Global Frequency Allocations

When radio frequency bands in the electromagnetic field are shared and managed between countries, this is called global frequency sharing. These assignments are very important to make sure that the spectrum can be used efficiently and without interference for many things, such as communication, television, tracking, scientific study, and more. The International Telecommunication Union (ITU), a specialized UN body, is in charge of making sure that frequencies are shared evenly around the world. The International Telecommunication Union (ITU) sets rules for how airwaves can be used around the world through international laws and deals.

Some important parts of world frequency assignments are:

❖ **Frequency Bands:** The spectrum is split up into different frequency bands, and each one is used for a different purpose, such as radio programming, mobile communications, satellite communications, flight, amateur radio, and so on.

❖ **International Harmonization:** To make sure that all devices can talk to each other and to reduce disturbance, international deals set aside certain frequency bands for different services or uses. For example, 5G cell networks only work within certain frequency bands so that they can work in all countries.

❖ **Regional and National Allocations:** Global allocations set the rules, but each country is responsible for controlling how its airwaves are used within its limits. Frequencies are assigned by national regulatory bodies based on the needs of each country, but these assignments are usually in line with global standards to make it easier for people to talk to each other across borders.

❖ **Coordinating Frequencies:** It is important for nearby countries, especially those that share borders, to work together to handle common frequencies and keep disturbance from happening. To make this collaboration easier, deals can be made between two or more countries.

- ❖ **Spectrum bids and Licensing**: Frequency bands are often given to telecommunications companies, media, and other businesses by the government through bids or licensing processes.
- ❖ **New Technologies**: As technology improves, the need for certain frequency bands may change. The ITU changes assignments every so often to make room for new technologies while still making sure that old system can still work with them.
- ❖ **Non-Interference Principle**: To keep things running smoothly, international deals stress the need for radio services that share the same frequency band to not interact with each other.

Spectrum management is the process of making sure that different people can live without interfering with each other. This is done by understanding world frequency licenses. Following these rules makes it easier for people to talk to each other, for new technologies to be developed, and for the radio frequency band to be used efficiently across countries.

Understanding ITU Radio Regulations

The International Telecommunication Union (ITU) Radio Regulations are a large international agreement that controls how radio waves are used and how satellite paths are managed. They are rules and standards made by the ITU, a special body of the UN, to make sure that the radio-frequency band is used fairly and efficiently by all radiocommunication services around the world.

Background and Purpose of ITU Radio Regulations

Spectrum Allocation and Management

- The radio-frequency band is a limited and valuable resource that is used for many types of communication, such as phone calls, broadcasts, satellite communications, and more. The ITU Radio Regulations are very important for handling global spectrum sharing, giving frequency bands, and making sure that different radio services can work without conflict.

International Coordination

- The ITU Radio Regulations make it easier for countries and telecommunications organizations to work together internationally because the need for wireless services is growing around the world. These laws help in settling disputes related to frequency sharing and satellite paths, supporting peaceful living among different wireless technologies and services.

Framework for Innovation and Development

- The ITU Radio Regulations provide a safe structure that promotes innovation and the development of new wireless technologies. Creating uniform rules and procedures promotes a favorable environment for technological advances and the launch of new radiocommunication systems.

Key Components of ITU Radio Regulations

- ❖ **Frequency Allocation**: The laws group and allocate specific frequency bands for different services and uses. These amounts are based on international deals and are periodically updated to suit changing technologies and demands.
- ❖ **Radio Station Licensing**: The ITU Radio Regulations set out processes and standards for licensing radio stations. This includes the conditions for getting permits, technical specs, and working factors to ensure agreement with foreign standards.
- ❖ **Spectrum Sharing and Interference** Management: To avoid interference between different radio services running in neighboring frequency bands, the rules describe methods for spectrum sharing and interference reduction. This includes cooperation between nearby countries and services to reduce delays and keep efficient radio use.
- ❖ **Satellite Orbit Management**: The ITU Radio Regulations also control the use and coordination of satellite orbits and orders. They create processes for registering satellite networks, giving orbital places, and ensuring the reasonable and fair use of the geostationary and non-geostationary satellite orbits.
- ❖ **International Radio Communication Conference (WRC):** The laws are regularly reviewed and changed through the World Radiocommunication Conferences (WRCs) organized by the ITU. These workshops gather officials from member states to discuss and rewrite the Radio Regulations to suit technological advances and solve new challenges.

Significance and Impact

❖ **Global Harmonization**: The ITU Radio Regulations enable global harmonization of radio-frequency bandwidth control. This standardization ensures interoperability among different radio systems, allowing smooth contact services across countries.

❖ **Spectrum Efficiency**: Efficient spectrum management allowed by these laws supports efficient usage of the radio-frequency band. It enables the growth of wireless services and devices while reducing disturbance and frequency waste.

❖ **International Cooperation:** The ITU Radio Regulations encourage countries, telecommunications companies, and other industry players to work together by setting up a framework for international cooperation and planning. This cooperation is very important for solving global spectrum problems and making sure that everyone has equal access to the airwaves.

❖ **Technological Innovation**: By offering a safe and uniform framework, the regulations promote innovation in the area of radio transmission. This encourages the creation and use of new technologies, which is good for businesses and people all over the world.

Challenges and Evolving Landscape

Spectrum Scarcity

- There are worries about bandwidth shortage because there are so many wireless devices and more people want to use spectrum-intensive apps (5G, IoT, etc.). It will be hard for the ITU Radio Regulations to keep up with these growing needs while still allowing for efficient use of the airwaves.

Emerging Technologies

- The arrival of new technologies like cognitive radio and dynamic spectrum access makes it harder to use old ways of managing the spectrum. The regulations still face a major task in adjusting to these technologies while still protecting against meddling.

Regulatory Adaptation

- The rate of technological progress is often faster than the rate at which rules and regulations can change. The ITU Radio Regulations need to be constantly revised and updated so that they can keep up with new technologies and changing phone networks. In conclusion, the ITU Radio Regulations are an important set of rules for managing the global spectrum. They make it easier for countries to work together, make good use of the spectrum, and come up with new radio communication technologies. To address new challenges and meet the world's ever-increasing demand for wireless communication services, these regulations must constantly evolve and change.

Harmonizing Baofeng Radios for International Use

Harmonizing Baofeng radios for use around the world means taking care of a few important issues to make sure they work with radio rules and regulations in every country. Amateur radio users and fans all over the world love Baofeng radios because they are affordable and can be used in a variety of situations. However, because different countries have different frequency bands, power levels, and licensing standards, it may be hard for these radios to meet the rules in different countries. When making Baofeng radios work with other radios around the world, things like

Frequency Bands and Compliance

Regional Frequency Allocation

- Radio frequency bands are assigned differently in various countries and regions. To meet local regulations, Baofeng radios must support the frequency bands set aside for amateur radio use in particular nations or regions.
- Knowing the amateur radio bands and allowed frequency ranges in different regions (for example, 2 meters and 70 centimeters) is important for setting Baofeng radios to work properly in those areas.

Coverage of Frequency Range

- Both amateur radio bands and other frequencies are frequently covered by Baofeng radios, which have a wide frequency range. Users must make sure they follow the rules by only sending signals to amateur bands that are allowed in their country.

Transmit Power Levels
Power Output Limitations

- The highest power level for amateur radios is regulated by radio regulations in various countries. To avoid going over the legal limits in particular regions, Baofeng radios may need to be set or changed to work with these power limits.

Type Approval and Certification
Regulatory Compliance and Certification

- Individual countries or regions have rules about how Baofeng radios must be certified and work. For legal use, the gadget must meet local certification standards. For example, in the US, it must have FCC certification, and in Europe, it must have CE marking.

Compliance Programming and Modification
Programming and Locking

- Users should know how to set Baofeng radios to work only in the allowed frequency ranges and power levels, blocking frequencies that aren't allowed. To set limits in accordance with local regulations, computer tools and processes must be used.
- Some Baofeng radios have a feature that lets you lock frequency ranges or power settings to make sure they are followed and stop people from not following the rules on purpose or by accident.

User Education and Awareness
Regulatory Awareness

- If amateur radio users plan to use Baofeng radios in other countries, they should make sure they know the rules and regulations for radios in those countries. For compliance, it is very important to know about area rules, allowed bands, and power limits.

CHAPTER 16
MAINTENANCE AND TROUBLESHOOTING
Care and Maintenance Tips
Cleaning and Handling the Radio

To keep your Baofeng radio working well and looking good, you need to clean it and handle it the right way. Here is a full guide on how to clean and take care of your Baofeng radio:

Cleaning

1. External Cleaning:
- Wipe down the outside of the radio often with a soft, dry cloth to get rid of dust, dirt, or fingerprints.
- To get rid of tough spots or dirt, dampen the cloth slightly with water or a mild soap solution (but don't let it drip wet) and wipe the surface gently. Do not let water get into the ports or holes.
- Don't use rough chemicals, cleaners, or materials that can scratch the radio's shine or logos.

2. Display and buttons:
- Clean a makeup brush or a small, soft toothbrush with soft bristles. 2. Use this brush to remove dust or dirt from the screen and around the buttons.
- To clean hard-to-reach places, lightly wet the brush with water or isopropyl alcohol (70% strength or less). Make sure that the brush is not wet.

3. The antenna:
- Check the antenna often to see if dirt, dust, or grime has built up. To clean it gently, use a soft cloth or a cotton swab that has been wet with water or a light soap solution. Do not scrub too hard.

4. Battery and Contacts:
- Take the battery out of the radio before you clean the wires or battery.
- Clean the battery contacts and the radio contacts that go with them with a dry cloth or cotton swab. Make sure they don't have any dust, dirt, or other residue on them that could get in the way of the connection.

5. Drying:

- Make sure the radio is fully dry after cleaning it before putting it back together or using it. If the radio gets wet, let it dry naturally or use a dry cloth to wipe off any extra water.

Handling

1. Avoid Dropping:
   - Be careful when handling the radio and don't drop it, as the impact could damage the insides or the outside shell.

2. Proper Storage:
   - Keep the radio in a cool, dry place that is out of direct sunlight and away from rough weather. To keep it from getting damaged by wetness, don't store it in damp places.

3. Use the right accessories:
   - Always use the chargers, batteries, radios, and other tools that work with your device that the maker suggests. Using devices that don't work with the radio can damage it.

4. Avoid Excessive Force:
   - Be careful when you turn knobs or press buttons. Using too much force can damage or wear out the buttons faster than they should.

5. Temperature Consideration:
   - Keep the radio away from very hot or very cold weather; doing so can damage the radio's parts and shorten the battery life.

6. Transportation:
   - Use a bag or pouch to protect the radio when you carry it to keep it from getting scratched or damaged. Make sure it's locked down so it doesn't fall off by chance.

Battery Care and Replacement

Battery Care

∞ First Charging: Fully charge the battery before using the radio for the first time after getting it. Follow the manufacturer's instructions for how long to charge the device for the first time.

∞ Use compatible chargers: Always use a charger that is made to work with the battery in your Baofeng radio. If you use the wrong chargers, you could hurt the battery.

- ∞ **Don't charge the battery too much**; this can hurt the battery. As soon as the battery is fully charged, take it off the charger.
- ∞ **Store the battery correctly**: If you're not going to use it for a while, put it somewhere cool and dry with about half of its charge still on it. This helps it last a long time.
- ∞ **Stay away from extreme temperatures**. Batteries can be hurt by high temps. Do not leave the radio out in full sunlight or places that are very hot for long amounts of time.
- ∞ **Regular Use**: To keep the battery in good shape, use it and charge it regularly. Don't let it flow for long amounts of time.

Battery Replacement

- ❖ **Make sure the battery is healthy**. If the battery life has significantly decreased or if it doesn't hold a charge well, it may be time to get a new one.
- ❖ **Buy Original Batteries**: To make sure they work and are of good quality, you should only buy new batteries from authorized sellers or the maker.
- ❖ **Follow Instructions**: When changing the battery, do what the manufacturer or user manual says. Before taking out or putting in the battery, make sure the radio is off.
- ❖ **Correct Installation**: Put the new battery in the right way and firmly. Damage to the radio or the new battery could happen if the fitting is done wrong.
- ❖ **Dispose of Old Batteries Properly**: If the old battery isn't working anymore, get rid of it the right way in accordance with local regulations. A lot of shops that sell electronics or recycle sites will take old batteries and get rid of them properly.
- ❖ **Testing**: Once you've changed the battery, make sure the radio works properly by testing it.

Troubleshooting Common Issues

Baofeng UV-82 Troubleshooting

The most common problems with the radio and how to fix them are on the list below.

1. **Battery won't hold a charge**

A Baofeng UV-82 battery that won't hold a charge is one of the items that break down the most. There are a few things that could cause this, like using a bad or broken battery or not charging the cell properly. To figure out what's wrong, make sure the battery is fully charged and that it is properly placed on the radio. If that doesn't fix the problem, try a different battery. If you're not using the original charger, it might not be as good and could be what's wrong.

2. Poor transmission or reception

Poor transfer or reception is another problem that a lot of people have with the Baofeng UV-82. Having a bad receiver or being on the wrong frequency are just two of the things that could cause this. To figure out what's wrong, make sure the antenna is properly attached to the radio and that it works properly. Make sure the radio is set to the right frequency for the type of broadcast or response you want. If you need to, change the mute setting; it helps cut down on background noise. If you are still having trouble with transfer or reception, try moving to a different area or calling the maker for more help.

3. Keypad not responding

Users of the Baofeng UV-82 may find it annoying when the keyboard doesn't work. This could be due to some things, like a keyboard that doesn't work or is locked. Make sure the passcode lock is not in place and try setting the radio back to its original settings to see if that fixes the problem. It might be necessary to contact the maker for more help if the keyboard still won't work.

4. Audio that isn't clear

Audio that is distorted can make it hard to understand what is being said. This can be caused by a broken speaker or too high of a volume setting, among other things. To fix this problem, make sure the speaker connections are tight and the volume isn't turned up too high. Check to see if the distortion goes away after you turn the sound up or down to a comfortable level. If the problem keeps happening, you should call the maker for more help.

5. Inability to program channels

Setting up channels can be tricky, and problems do happen from time to time. This might happen if you use the wrong programming wire or aren't in the right programming mode, among other things. Make sure the radio is in the right programming mode and that the right programming wire is being used to fix the problem. If the problem persists, look at the user instructions or get in touch with the maker for more help.

It can be hard to figure out how to change the settings on your Baofeng UV-82 if you aren't familiar with the radio's menu and settings. To fix this problem, look at the user manual to learn how to use the radio's menu and change its settings. Check to see if the radio is in the right mode for the settings you want. For more help, if you're still having problems, you should call the maker. Good luck with that! The second one isn't a problem by itself, but not knowing how to use the radio might be. On this radio, the different settings are hard to figure out if you haven't used this kind of radio before. To sum up, the Baofeng UV-82 is a durable and flexible radio, but it can have problems like any other tech device. If you know what the most common problems are and how to fix them, you can quickly get your radios working again. We know that the user guide isn't always clear and that it can be hard to find the answers you need. You should try to get to know your radio. You will learn more about how it works if you use it over and over again. This will help you find an answer much more quickly if something goes wrong.

Troubleshooting Common Issues on Baofeng Radio Programming

Programming Baofeng radios with CHIRP is usually pretty easy, but you might run into some problems. Here are some usual problems and how to fix them:

❖ **Connection Errors**: Make sure the programming wire is properly attached and that you have chosen the right COM port in CHIRP.
❖ **Invalid Frequency Error**: Check that the bands you're setting are legal in your area and within the range of your Baofeng radio.
❖ **Radio Not Responding**: To get into programming mode on some Baofeng types, you have to press certain buttons in a certain order. For the right way to do something, look at the user instructions or online tools.

Seven Steps to Avoid Most Baofeng Radio Problems

- **First**, none of the cheaper Chinese radios (and much cost a lot more than the Baofeng series) can be programmed the same way as the more expensive Japanese radios, either from the radio's keyboard or from free programming tools. If you look at the LCD of a Baofeng radio and see the "+-" offset sign, that doesn't mean that the radio has already set the right offset for the band and memory you are using. Additionally, the S-meter on Baofeng handhelds will always show full scale, no matter how strong the signal is. This can lead many users to believe their signal is better than it is. This is true for all Baofeng handheld products.

- **Second**, if you want to buy a Baofeng radio and think that FCC Part 90 approval also means that you can have two VHF or UHF frequencies 12.5kHz apart (or less) on your two VFOs and not hear any cross-channel noise or interference on the other frequency, you will need to buy a radio that is better than the Baofeng UV-5R. Many foreign sellers on Amazon or eBay don't care to ship FCC Part 90 radios because non-FCC Part 90 radios are usually cheaper. This means that these radios can't be used on any business frequency in the U.S. (*see picture below*). The radio on the right passed the QC test while the radio on the left failed.

The UV-5R is a cheap option, but if you want a better Baofeng receiver and can use many of the accessories that work with the UV-5R family, I suggest you look at the UV-5R V2+ (released in late 2014), the UV-5R 8 Watt, or the BF-F9V2+. They don't cost much more and have better Baofeng receiver boards. If you don't need all UV-5R accessories to work with your camera, check out the UV-82 line. If you're willing to spend a little more, both TYT and Wouxun make radios with better receivers, more memories (from 199 to 999), and a huge selection of accessories. TYT also has some features that Baofeng doesn't have, such as 10 watts of transmit power, 20 to 25 FM commercial radio channels, a multichannel audio circuit, and audio inversion for safer communication when it's available.

- **Third**, know that the Baofeng radios' **Twin Dual Receive (TDR) feature [Menu 7]** does NOT mean that you can listen to two different talks at the same time. The radio only scans back and forth between the two VFO receivers (think of counting "1001 and 1002" to get an idea of how fast it scans), so you might miss a short chat or comment on one VFO every once in a while. You can't hear two conversations at the same time on any current Baofeng radios, and you can't make them work that way with third-party "mods" or by buying a better radio.

- **Fourth**, if you insist on buying the cheapest USB programming cable, you should know that most new owners don't read or follow the detailed step-by-step instructions on the different internet radio forums on how to fix problems with cheap fake cables and/or they don't know how to turn off the Wind properly.

Also, you will need to use radio programming software to use the radio's alpha-numeric memory naming. You can't do it with the keyboard on any Baofeng radio, but you can do it with many TYT radio types. If you want to fully escape the cord issue, I recommend getting a real Prolific, Silicon Labs, or FTDI 2-prong Kenwood-style programming cable. Most of the problems people have with programming their Baofeng radios can be fixed by buying a real (not fake) cable and following the instructions on online forums. You could also take your programming to the next level and get an RT Systems cable and software kit, which will have your radio programmed faster than you can drink your first cup of coffee. We use and sell RT Systems kits. Having people who: *a) speak English as their first language; b) are U.S. ham radio operators; c) have your model radio in front of them; and d) support their product* is worth every penny, but I know that some people will still want to waste a few hours of their lives by trying their fake $5.00 programming cables and freeware first.

Three main companies make programming data cables: FTDI, Prolific, and Silicon Labs. The Prolific cables are the ones that are illegally copied the most from China, so you never know what you'll get. A real Prolific, Silicon Labs, or FTDI wire will cost between $20 and $25, but it will save you hours of hair-pulling, which some of us must have. The RT Systems kit with the FTDI cable and radio software that goes with it costs $48.95, but your friends will think you're a Baofeng programming radio "genius" in no time. If you want to use free programming tools instead, like Chirp or VIP, you should get either an FTDI or Silicon Labs cable. The Wouxun/Red Silicon Labs programming cable is my favorite, but FTDI cords also work well. Again, if you don't want to deal with programming (and getting the USB cable driver to work), or if you think your time is still worth something, call us about the different Baofeng software cable kits from RT Systems. They cost more than a fake Prolific cable with free software, but they work, and they work very well. In some places, the bragging rights alone might be worth the $48.95. You can keep it a secret how you deal with being called a computer "*genius*" because you use the RT Systems kit. Also, we sell software and wire kits for a huge number of radios from other brands.

- **Fifth**, most of the complaints about this brand of radios (and Chinese radios in general) will come from older hams or engineers who keep saying that "they" should be able to do things with it (using a different method) and can't figure out why their $60 to $200 radio doesn't work like their $300 to $600 Japanese radios from Alinco, Kenwood, Icom, or Yaesu.

People then complain about Chinese radios on different sites, mostly because the owners won't follow directions or the radio doesn't work the way "they" think it should, or because it doesn't have all the features of their $300 to $600 Japanese radio, such as APRS, D-Star, or GPS. People who have these problems also don't know why their friends who use PL tones or Privacy Codes on their different FRS/GMRS radios don't automatically hear them when they send on the right frequency (but without a PL tone on for transmission). People also often don't realize that later digital trunking systems can't be talked to or heard with any analog radio, not just the different Baofeng radios. You'll need to buy either a digital radio or a digital scanner if you want to keep an eye on any local police department that has gone digital. Luckily for us, the Arizona Department of Public Safety and many county sheriffs still use traditional VHF or UHF. All dual-band Chinese radios can easily be set to use this frequency (as long as you have the right wire and software).

- **Sixth**, know this: if you don't know much about newer VHF/UHF radios or don't want to follow the step-by-step steps you can find on different Baofeng sites or user groups, or if you don't even plan to read the manual, you will probably have trouble with these radios at some point. Most of the time, it won't be the radio's fault because you either won't change the frequency step properly or will set the PL tone on both encode and decode, (thereby blocking the repeater audio from coming through to your receiver), etc., etc., but you'll blame the radio nonetheless.

I think you should buy a radio that has already been "*pre-programmed*" with many of the bands in your area. This will depend on the features and memory channels of your radio, which can be different from one model to the next. We can set the radio with a common script for just $20, which will save you a lot of trouble and give you a printout of the radio's frequency and channel.

- *Seventh*, the people who sell radios are all over the place, but I'd recommend finding one who is at least based in the U.S. and can give you the "*straight scoop*" on the different Baofeng types. It's a given that sellers who only offer one brand of radio won't be able to help you find a better one for your needs, since they don't carry that other brand or model.

Several Chinese sellers also say they are based in the U.S. to get "*local*" business, or they say they have direct contact with the Chinese maker and even say they are the only ones who can fix a certain brand of radio, which is absolute nonsense. It's also important to know that some sellers will take "seconds" from the plant, which could be radios that have a problem that is important to you! We insist on only getting brand-new radios, not rejects, seconds, or units that have been sent back before. We also insist on FCC Part 90 radios when they come in (as we already said, many of the ones sold on the internet aren't meant to be sold in the U.S.). If you want to take a chance on one of those types, there are many Amazon and eBay sellers (mostly from China) who will gladly take your money but will make you pay to send back the broken radio before they give you your money back.

Finding out how much it costs to send your broken radio back to China will surprise you. (As was already said, a lot of radios aren't broken; they're just not set right, and the hardware of the radio is blamed for the problem.) You did great if you made it this far. If you follow these seven easy steps, you will be able to avoid most of the setting problems that new Baofeng owners have. As a result, you will be a smarter radio buyer.

Frequently Asked Questions

What is a Baofeng?

It's a brand name for a very popular make of cheap handheld radio, favored by radio amateurs and other radio enthusiasts.

Can I use it for amateur radio?

Yes, as long as you have a license. A lot of beginners use Baofeng because they are cheap and strong. On a Baofeng, it is allowed to listen to amateur radio but to send, you need a current amateur radio license from Ofcom. To do this, you need to get a "Foundation" license for amateur radio in the UK. Most of the time, this means studying for six hours, doing some simple practice tasks, and taking a multiple-choice theory test. You can learn online or with the help of an amateur radio club in your area. You have to pay £27.50 to take the test, but the Ofcom license is free. In the amateur radio frequency ranges of 144 to 146MHz and 430 to 440MHz, you can use a small radio once you have your license. Both of these areas can be reached by radios like the well-known Baofeng UV-5R. It is not against the law to use a Baofeng to listen to amateur radio without a license.

Can I use it for PMR 446MHz?

Legally, no.

Consumer short-range "**walkie-talkie**" handhelds without a license are generally referred to as "PMR" (Private Mobile Radio) and operate in the frequency band 446.0 to 446.2MHz in the United Kingdom. There is no need for a license to use radio technology as long as it meets certain technical standards. Technically, Baofengs can be set up to work on the 446MHz UK PMR channels. However, it is against the law to use a Baofeng to send on those frequencies. This is why: It is illegal for PMR446 devices to send more than 0.5 watts of power. Most Baofeng radios send at 5 watts or 8 watts. Some Baofengs do have a low-power setting, but it's at least 1 watt, which means that even when it's set to low power, it sends more power than if it were set to PMR446 in the UK. Many people say that sending 1 watt instead of 0.5 watts won't be noticed or cared about, but they do so at their own risk. The official answer, as we understand it, is "No, you can't legally use a Baofeng on 446MHz if it is sending more than 0.5 watts." There is no law against listening to PMR446 on a Baofeng mobile, though. Keep in mind that there are other rules about the gear that can be used for PMR446. For example, the radio must have a fixed antenna that can't be taken off, and it must also meet certain technical standards outlined in the Ofcom advice documents (as shown by the CE Mark).

Can I use it for Ofcom Business Licence?

The Ofcom Business Simple UK Licence is another way to properly use Baofeng radios if you don't want an amateur radio license. This can be bought online from Ofcom for £75 for 5 years at the time of writing. It is possible to get a license as a person, a business, a group, or a charity. You can use a set of bands between 164MHz and 449MHz with this license. This license only lets you use 5 watts of power, which is the normal amount of power for most Baofeng types.

Are there any frequencies I can transmit without a license?

No. Not legally.

Which Baofeng radio is the best?

Of course, the best Baofeng radio for you will rely on your wants and tastes, but a genuine Baofeng model is a good place to start. Once you know the brand, there are a few other things you should think about before making your choice.

Does Baofeng receive frequencies from aircraft?

When it comes to getting bands from planes, Baofeng radios can't handle them. Usually, airplanes use AM frequencies, but Baofeng radios are made to work with FM frequencies.

How much do Baofeng radios cost?

People like Baofeng radios because they are known for being cheap. You can get these radios for pretty cheap; a good one starts at as little as $30. This makes them a good choice for people who want reliable contact tools without spending a lot of money.

Can I make my Baofeng radio's range better?

You can make your Baofeng radio's range longer by connecting an extra antenna. This can improve both receiving and communication compared to the antenna that comes with the radio.

Does Baofeng radio come with a warranty?

Most Baofeng radios come with a guarantee and customer service, so buyers know what to do if there are any problems with the product or problems with the way it was made.

How many Watts do Baofeng radios have?

Baofeng radios usually have wattages between 1 and 8 watts, so they can output different amounts of power to meet different contact needs.

Can I use the radio for general communication with family?

For general family or group contact, these radios can be used. However, it's important to remember that you usually need to get a license to properly use them for this.

Can I use the radio as a scanner?

Baofeng radios can work as scanners, which let users quickly switch between different bands and channels while listening to different broadcasts.

How far does the Baofeng UV 82 work?

The transmittal range usually changes based on the situation. When things are normal, a Baofeng UV 82 can send signals up to 10 miles away. Again, it can be sent up to two miles away when conditions are bad. A Baofeng UV 82, on the other hand, can send light up to 15 miles away when conditions are perfect.

Conclusion

The conclusion is that Baofeng radios provide a communication solution that is both economical and adaptable, making them suitable for a wide range of settings, from outdoor activities to emergencies. An overview of the features, functions, and fundamental operations of Baofeng radios has been presented in this guide. The purpose of this guide is to aid users in making the most of the possibilities that these devices have to offer. Baofeng radios are dependable companions for communication in a variety of environments due to their extensive frequency range, customizable options, and small shape. The safe and legal use of these radios is ensured by having an understanding of the programming techniques, the significance of obtaining the appropriate license, and the relevance of adhering to the local legislation. By adhering to the standards that are stated in this guide and continually exploring the features of Baofeng radios, users can fully harness the potential of these radios for successful communication in a variety of situations.

INDEX

1

10 X Original Antenna for BaoFeng Bf-888s, 109
12V cigarette plugs, 19, 28
16 Channels, 8

2

2 Pack-29 Inches Foldable/Tactical Radio Antenna, 106

A

A bad connection, 104
A Baofeng programming cable, 58
A Baofeng radio's body, 19
A Guide to Saving Your Current Configuration, 49
A high front-to-back ratio, 118
A pair of Baofeng/Pofung series transceivers, 91
ABBREE AR-152A, 26
Absorption properties of different materials, 127
Accessibility and Comfort Considerations, 133
Accessible Charging Ports, 134
accessing the GPS information, 96
Accessories and Compatibility, 11
Accidental Damage, 56
Accuracy and Precision, 46
Acronym, 41
Active devices for signal intensification, 123
Active devices for signal intensification are being added., 123
Actual-time Scanning and Monitoring, 5
add GPS and APRS, 95
Add to scanner list, 40
Added Considerations for Field Operation, 92
adding new technologies, 98
Additional Tips and Best Practices, 56
Additional Tips and Best Practices for Optimal Baofeng Radio, 56
Additional Tips and Best Practices for Optimal Baofeng Radio Performance, 56
Adjustable Settings, 134

Adjusting Volume and Squelch, 32
ADVANCED FEATURES AND FUNCTIONS, 89
Air Travel Rules, 5
Aligning the Battery, 28
Amateur Radio, 90, 93
Amateur Radio (Ham Radio) Use, 90
amateur radio fans, 5, 13
amateur radio fans to professionals working in many fields., 5
amateur radio users, 97, 98
ANI Code box, 62
ANI data burst, 63
Antenna Essential for Transmission, 27
Antenna Matters, 57
Antenna Protection, 132
Antenna Protector, 132
Antenna Separation, 92
Antenna Upgrades, 104
Antenna-Best long-range antennas, 105
Antennas for baofeng 5R And BF-888s Radio, 105
APPLICATION, 101
Apply for a new license, 100
appropriate frequency ranges, 18
appropriate license, 154
APRS, 95, 96, 97, 98, 150
APRS TNC, 95, 96
Are there any frequencies I can transmit without a license?, 153
Assigning Channels to Groups, 70
Attaching the Cable, 39
Audio that isn't clear, 146
Auditory Feedback, 134
Authentic Genuine Nagoya, 105
Authorizations link, 102
Automatic Number Identification (ANI), 62
Automatic Packet Reporting System, 95, 97
automatic units, 35
avoid any electricity problems, 29
Avoid Dropping, 144
Avoid Excessive Force, 144
Avoid Interference, 5
avoid interruptions, 23

B

background noise, 9, 32, 33, 67, 146
Backing up Radio Configuration, 48
Backlit LCD, 14
Backlit LCD Display, 14
Backup setups, 58
Band, 4, 10, 12, 14, 22, 40, 89, 90, 92, 99, 105, 107, 108, 110, 111, 112, 113, 127
Bands and Compliance, 141
Baofeng antenna, 104, 114
Baofeng BF-888S, 8, 9, 109
BaoFeng bf-888s, 105, 109
BaoFeng bf-888s Antenna, 105, 109
Baofeng BF-F8HP, 13, 108
Baofeng GMRS-V1, 81
Baofeng GT-3TP, 11, 12
Baofeng handheld products, 147
Baofeng Magnetic Antenna, 19, 26
Baofeng Magnetic Car Vehicle, 105, 111
Baofeng Magnetic Car Vehicle Mounted, 105, 111
Baofeng models, 4, 14, 114
Baofeng radio clockwise, 32
BAOFENG RADIO COMPONENTS, 19
Baofeng radio models, 4
Baofeng radios and CHIRP software, 42
Baofeng radios are dependable companions for communication, 154
Baofeng Radios Use Cases and Applications, 14
Baofeng software, 39, 65, 149
Baofeng Software, 39
BaoFeng SRH805S, 105, 111
BaoFeng SRH805S SMA-F, 105, 111
BaoFeng SRH805S SMA-F Female, 105
Baofeng transceiver's lower jack, 91
Baofeng transceiver's top jack, 91
Baofeng types, 12, 15, 147, 151, 153
baofeng UV-5r, 105
Baofeng UV-5R, 6, 7, 13, 65, 70, 71, 81, 105, 111, 148, 152
Baofeng UV-5R radios, 65, 71
Baofeng UV-5R series, 6, 7, 13
Baofeng UV-82, 9, 10, 11, 24, 58, 61, 62, 66, 71, 85, 87, 88, 107, 110, 145, 146
Baofeng's well-known GT-3 series, 11
Basic Controls and Functions, 32
basic guideline for integrating GPS and APRS functionality with a Baofeng radio, 95

Battery, 4, 7, 8, 10, 14, 19, 24, 28, 30, 31, 57, 68, 132, 134, 135, 143, 144, 145
Battery and Charging Options, 19
Battery Level Indicator, 24
Battery Life, 7, 8, 10, 134
battery pack, 30, 57, 74
battery's contact plates, 29
Be careful with airbags, 5
beamwidth, 118, 119
Beamwidth, 118
Belt Clip, 27
Belt clips or holsters, 135
Best Antennas For baofenG-82, 105
best Baofeng antenna, 104, 112, 114
Better Reception Quality, 117
Better Signal Strength, 116
black bars, 24
Block Settings, 57
block unwanted signals, 41
blocked frequencies, 53, 54, 57
Blocking, 42, 43, 52, 53, 54
Bob Bruninga, 97
BOTH, 63
Braille Labels or Tactile Lines, 135
brand of radio, 42, 151
brief overview of some common digital modes, 93
Broader Channel Capabilities, 4
built-in flashlight features, 4
Built-in flashlight features, 4
business radio, 17
businessperson, 42
BusyLock, 40
But what's the difference and what does your company need?, 16
buttons and settings, 12, 135
buy high-quality antennas, 57
Buy Original Batteries, 145
Buying scanner, 51
Buying scanner and spectrum analyzer tools, 51

C

C4FM, 93
Cable Integrity, 48
Calling a User Group, 72
callsigns, 84
Can I make my Baofeng radio's range better?, 153
Can I use it for Ofcom Business Licence?, 152

Can I use it for PMR 446MHz?, 152
Can I use the radio as a scanner?, 154
Can I use the radio for general communication with family?, 154
Capabilities, 89
care and maintenance for your radio, 31
Care and Maintenance Tips, 143
Care and Replacement, 144
Care when handling, 27
Carrier Operation, 85
Carrying Case or Holster, 132
CBRS, 99
cellular network coverage, 15
Challenges and Evolving Landscape, 140
Change and Upload, 47
change bands, 77
change frequencies, 41, 47, 57
Change the radio's operating band, 77
change the reader setting, 85
Change the setup, 56
changing electronic devices, 135
Changing the antenna's height and direction, 123
Changing the communication medium, 128
Channel, 4, 7, 40, 53, 57, 70, 72, 74, 76, 78, 85, 88
channel bandwidth, 81
Channel Information, 40
Channel Information Window, 40
Channel Memory, 7
channel mode, 21, 24, 35, 37, 65, 87
Channel number., 40
channels and frequencies, 5, 7, 33
charge a wet battery, 31
charge Baofeng radios, 19
Chargers, 28
Charging Accessories, 135
Charging the Battery, 28
Check and Test, 57
Checking Included Accessories, 26
CHIRP, 41, 42, 43, 44, 45, 46, 47, 48, 49, 50, 53, 54, 55, 56, 61, 80, 81, 85, 147
Chirp or VIP, 149
Chirp programming software, 58
CHIRP software, 42, 49, 80
CH-Name, 40
choice of antenna polarization, 127, 128
Choose a language, 39
Choose Frequencies to Block, 53
Choose the Right COM Port, 48

CHOOSING THE RIGHT PROTECTIVE GEAR, 132
Circular Polarization, 126
Cleaning, 143
Cleaning and Handling the Radio, 143
coaxial connection, 111
Collaboration with Regulatory Authorities, 51
COM port, 47, 48, 147
Comfort and Wearability, 134
Comfortable Ways to Carry, 134
commercial transceivers, 103
common link problems,, 48
Common Pitfalls and Best Practices, 70
Comms Ham Radio Interoperability Programming, 41
communication experience, 13
Communication Options, 89
communication pros., 13
Communication Protocol, 47
Community and Support, 11
Compact and Portable Design, 4
Compact Design, 8, 11
Compatibility, 14, 70, 94, 114
Compatibility with Accessories, 14
Compatible with Accessories, 13
Complete Removal, 30
Compliance with Accessibility Standards, 135
Compromised Security, 43
Computer Programming, 39
Conclusion, 154
Conducting Group Calls and Announcements, 72
Connect to Computer, 39
Connect via USB, 80
Connect Your Radio, 55
Connecting Baofeng Radio, 47
Connecting Your Radio, 44
Connection, 40, 47, 48, 91, 147
Connection Test, 40
connections for the feedline, 123
Connector Alignment, 27
Construction and Security, 15
construction sites, 15
CONTINUE FOR PAYMENT OPTIONS, 101
CONTINUE TO CERTIFY, 101
Continue with Plastic Card Payment, 102
conversation, 3, 8, 42, 46, 56, 67, 68, 69, 87, 90, 95, 104, 117, 130
Coordinating Frequencies, 137
CORES, 99
Correct Installation, 145

cost-effective choice, 8
Creating and Saving New Configuration, 55
cross-band repeat, 90
cross-platform program, 41
crowding and crosstalk, 16
CTCSS, 35, 37, 38, 40, 67, 68, 69, 70, 71, 73, 78, 79, 86
CTCSS or DCS, 67, 78
CTCSS tone, 35, 38, 68, 69, 78
current Baofeng radios, 149
current channels, 40
current licenses, 100
current settings, 46, 47, 49, 53, 80
Custom Cases or Grips, 135
Customer Support Accessibility, 134
customization options, 61

D

Damaged Antenna Warning, 5
Data Integrity, 46
Data Vulnerability, 43
DCS, 35, 40, 67, 68, 69, 70, 71, 78, 79, 86
DCS tone, 35, 78
Dealing with Water Exposure, 32
Decode box, 62
Desired coverage area, 127
Detailed Instructions, 53
Detailed Instructions on Adding Frequencies, 53
Different areas and countries, 17
different countries, 17, 141
different frequencies, 4, 8, 14, 35, 67, 89, 90, 98, 103
different frequency ranges, 7, 22
different radio types, 41
different scanning modes, 85
different situations, 7, 32, 129
Different types of antenna polarization, 126
different ways to use the scanner, 85
digital DNA, 48
Digital Mobile Radio, 93
Digital Modes and Encryption, 93
Digital Modes techniques, 93
Digital Smart Technologies, 93
digital technology, 3
direction of the offset, 37
directive gain, 119
Directive Gain, 119
Directivity, 117
Display and buttons, 143

Display and Buttons, 19
Display and Feedback, 134
display's backlight, 24
Dispose of Old Batteries Properly, 145
Disturbance and Disruption, 43
division multiplexing, 93
DMR, 4, 93, 94, 95
Documentation, 56, 57
Documentation and Compliance, 56
Does Baofeng radio come with a warranty?, 153
Does Baofeng receive frequencies from aircraft?, 153
Don't charge the battery too much, 145
Don't Use Glue, 27
Download Electronic Authorizations, 102
Download from Radio, 45, 47, 53, 55, 59
Download From Radio, 49
Download Radio Settings, 80
Downloading and setting up, 44
D-Star, 93, 94, 150
DTMF Encode, 62
Dual Reception, 87
dual watch, 11, 14
Dual Watch, 7, 11, 70, 71, 87, 88
Dual Watch and Dual Reception, 7, 11
Dual-Band Capability, 10
Dual-Band Functionality, 14
Dual-Band Operation, 4, 12
Dual-band radios, 90
Duplexers, 92, 93
Duplexers for Single Antenna Use, 92
Durability, 114
Durable Build, 9

E

earpieces, 4, 11
easier-to-press buttons, 3
Easy-to-Grip Design, 134
economical and adaptable, 154
Education and Training, 15
efficiency, 68, 114, 119
Efficiency, 49, 114, 117, 140
Efficiency in Replication, 49
Efficient and Consistent, 56
emergencies, 4, 11, 56, 154
Emergency Alarm, 9
Emergency Alert, 7
Emergency Codes and Procedures, 68

Emergency Communication, 70, 74
Emergency Preparedness, 14
emergency sound feature, 7
Emerging Technologies, 140
enable the ANI settings, 63
Enabling and Using GPS Features, 96
Enabling/Disabling/Configuring ANI Settings, 63
Encryption, 94, 95
End of Transmission, 63
Enhanced Features, 90
Enhanced Range, 116
ensure compliance and safety, 103
Environmental Conditions, 127
Environmental Protection, 132
EOT, 63
Event Management, 15
EVOLUTION OF BAOFENG RADIO, 3
EVOLUTION OF BAOFENG RADIO MODELS, 3
excessive amount of power, 18
expert responder., 43
Explaining the Process of Testing, 54
Exploration and Field Research, 15
exploring the features, 154
exploring the features of Baofeng radios, 154
exposure limits, 102, 103
Exposure Limits, 102
Extended Antenna Options, 133
Extended antennas, 135
External Antenna, 27, 133
External Antenna Usage, 27
External Cleaning, 143
External Speakers or mics, 135
Extra Batteries, 132

F

Factors Affecting Antenna Gain, 119
Factors affecting antenna polarization, 127
Family Communication, 15
family-run a GMRS station, 99
FCC, 18, 51, 80, 81, 84, 99, 101, 102, 142, 148, 151
FCC License Manager,, 99
feature-packed, 8
Federal Communications Commission, 18, 51, 102
Federal transmission Commission, 18
Find the accessory port and open it up, 39
Find the screws, 27
Finish the Process, 50

Firmware Updates, 94
First Charging, 144
Fits a Variety of Accessories, 4
Flexibility, 90
flexible, 3, 4, 5, 6, 7, 11, 12, 14, 28, 42, 49, 53, 67, 87,
    97, 98, 106, 147
FM broadcasts, 17
FM Radio Receiver, 7
FM radio receiving, 13
Follow Instructions, 145
Following the rules, 133
Framework for Innovation and Development, 139
Frequencies and Channels, 67
frequency assignments, 57
frequency bands, 12, 17, 18, 41, 57, 77, 89, 103, 137,
    138, 139, 141
Frequency Database, 51
frequency mode, 21, 33, 35, 36, 86
frequency range, 8, 16, 17, 34, 108, 109, 111, 112,
    113, 120, 141, 154
Frequency Range, 8, 10, 89, 141
Frequency Ranges, 17
Frequency Ranges and Regulations, 17
Frequently Asked Questions, 55, 151
FRN and password, 99, 101
Front-to-Back Ratio, 118

G

General Mobile Radio Service, 18, 99, 101
getting people's attention, 7
Getting regular firmware updates, 57
GETTING STARTED WITH BAOFENG RADIOS, 26
Global Harmonization, 140
GMRS, 18, 80, 81, 82, 84, 99, 100, 101, 102, 150
Going back to the Normal Screen, 72
GPS and APRS Integration, 95
GPS Unit, 95
GPS/Ranging, 96
ground-based stations, 126
Group Communication Strategies, 70
Guarding Privacy and Security, 42
Guide to Saving the Newly Programmed
    Configuration, 55
Guided Attachment, 28

H

ham, 17, 18, 23, 58, 89, 90, 95, 105, 108, 114, 119, 149

Ham Radio, 41, 90, 105, 110, 114

hand-held radios, 80

handle GMRS, 84

Handling, 143, 144

Hands-Free Operation, 132

hardware and software, 47, 53

Hardware Modifications, 135

Harmonizing Baofeng Radios, 141

Harmonizing Baofeng Radios for International Use, 141

headphones or phones, 3

Help and documentation, 134

High Contrast Display, 134

Higher efficiency, 118

higher frequency, 16, 17

High-Gain Antenna, 12

Hiking and Mountaineering, 15

hobbyist, 43, 93

Hold the Device and Place It, 72

Horizontal Polarization, 126

How far does the Baofeng UV 82 work?, 154

How to Block Frequencies on Baofeng Radio with CHIRP, 41

How to Change the Scanner Mode, 85

How to delete a channel, 39

How to determine the polarization of an antenna, 128

How to Install a Battery, 28

How to Make Batteries Last Longer, 31

How to Pick the Best Antenna Gain, 121

How to program baofeng UV-82, 58, 59

hydride batteries, 19, 28

HYS SMA-Female Handheld, 105, 113

I

Identifying and Resolving, 80

IDENTIFYING FREQUENCIES, 51

Identifying Frequencies to Block, 51

IDENTIFYING FREQUENCIES TO BLOCK, 51

Implementing Encryption for Secure Communication, 94

important details like frequency, 7, 14

important information, 19

important messages, 43, 67

important skills for controlling the noise of electromagnetic waves, 51

important things about the Baofeng BF-888S, 8

improves speed and safety., 15

improves the user experience, 10

Inability to program channels, 146

individualized approach, 21

Initial Charging, 31

Input Frequencies, 53

Installing Drivers, 48

Instant messaging, 97

Instant Restoration, 56

integrating GPS and APRS functionality, 95

Integrity of the signal, 42

intelligent move, 42

interesting journey, 3

Interference, 52, 54, 80, 117, 138, 139

INTERFERENCE AVOIDANCE, 80

Interference Mitigation, 52

International Cooperation, 140

International Coordination, 138

International Harmonization, 137

International Radio Communication Conference, 139

INTERNATIONAL RADIO REGULATIONS, 137

Interoperability, 41, 90

interruptions, 13, 130

INTRODUCTION, 1

Ionospheric conditions, 127

ITU, 137, 138, 139, 140, 141

J

JARL, 93

journey with Baofeng radios, 2

K

K5QHD, 21

Key Components, 139

Key components of APRS, 97

Key components of APRS include, 97

Key features, 12, 13, 106

Key Features, 107, 109, 110, 111, 112, 113

Keypad Lock, 24

Keypad not responding, 146

kind of protection., 42

L

Labeling and Identifying Groups, 70
Language and localization, 134
Large Battery Capacity, 14
Large Capacity Battery, 12
Lastly, the tone, 38
LCD screen, 10, 32, 65, 66, 96
LCD Screen, 10
Learning Curve, 90
LEGAL AND REGULATORY CONSIDERATIONS, 99
Legal Compliance, 11
Legal Considerations, 52, 70
Length ABBREE SMA-Female, 105, 108
Less disturbance, 69, 92
Let Go to Listen, 72
License and rule compliance, 133
license authorization, 102
Licensed Bands or Modes, 95
Licensing, 18, 52, 68, 99, 138, 139
Licensing Requirements, 18
Licensing Requirements (if applicable), 18
Limiting multipath fading, 130
Linux, 44
Listen to radio signals, 33
Localization Settings, 134
logistics, 15
Longer Battery Life, 4
lot of channels, 4, 7

M

Mac – read the radio, 59
macOS, 44
MAINTENANCE AND TROUBLESHOOTING, 143
Make sure the battery is healthy, 145
making radios, 3
Male SMA Connector, 27
Manuals and Guides, 133
manufacturing, 15, 16
Many Baofeng models, 5
many Baofeng radio users, 104
Maritime and Boating, 15
Markdown Table, 118
Measurement of Field Strength, 121
Measuring Antenna Gain, 120
MEM-CH, 36

MEM-CH menu item, 36
MEM-CH., 36
memory channel, 64, 70
menu button, 33, 75
methods for optimization, 123
Mini-Repeater, 90
Mistaken Positives, 54
mitigate interference, 52
Modify or add your parameters, 58
MONI, 24, 62
MONI button, 24
Monitoring for Local Regulations, 80
Monitoring FRS/GMRS Traffic, 80
Monitoring FRS/GMRS Traffic with a UV-5R, 80
multiple frequencies, 7, 11, 53
Multiple Functions, 13, 14
Multiple Functions and Features, 13, 14
My Licenses page, 102

N

Narrowband, 40
Narrower beamwidth, 118
Navigating the Spectrum, 51
network of digipeaters, 98
New Technologies, 138
NOAA weather broadcast, 75
non-FCC Part 90 radios, 148
normal temperatures., 31
Notable Features and Capabilities, 4
Note on Duplexer Usage, 93

O

Object Tracking, 97
OEM antenna, 107
Ofcom, 18, 151, 152
official CHIRP website, 44
Offset, 37
On the radio, press the Menu button., 63
One last thing and this is a big one, before we start writing., 81
Operations, 89
Optimizing Antenna Gain for Better Performance, 122
Optional Features box, 63
Optional Settings, 96
orientation of an antenna, 124, 127

outdoor activities, 1, 4, 11, 14, 33, 154
Outdoor Activities, 14
Output Power, 10

P

Parasitic Element Gain, 119
Personal Safety Gear, 133
Physical Accessibility, 133
Picking the Right Tones and Codes, 69
PL communications, 78
position of PTT-ID, 40
Position Reporting, 97
POUND, 24, 87
Pound Button, 24
power adapter, 28
Power Conservation, 92
power functions, 31
Power Levels, 103, 142
power off the radio, 31
Power On Message, 65
Power Output, 9, 12, 14, 142
Powering On and Off, 31
Power-On Message, 65
Practical applications, 130
Practical applications of antenna polarization, 130
Press 25 or move the cursor to SFT-D, 37
Press the "Menu" button, 32, 33
Press the [EXIT] button to leave the page., 85, 86, 88
press the [MENU] button, 73, 85, 86, 87
Press the Menu button, 37, 38, 39
press the Menu button., 36, 64, 65, 70
press the Menu key., 63, 70
press the VFO/MR button, 33
Press VFO/MR, 35, 37, 78
Privacy, 42, 69, 95, 150
Privacy Preservation, 42
private line, 69, 78
Procedure, 63, 86, 87, 88
process of charging multiple devices faster, 19, 28
Professional and Industrial Use, 15
Program the A and B settings separately., 65
Programming, 4, 10, 35, 37, 41, 46, 52, 53, 70, 90, 94, 96, 142, 147
Programming and managing blocked frequencies, 53
Programming and Software, 10
PROGRAMMING BAOFENG RADIOS, 41
PROGRAMMING BAOFENG RADIOS WITH CHIRP SOFTWARE, 41
Programming Blocked Frequencies, 52
PROGRAMMING CHANNELS, 35
programming cords, 7, 13, 14
Programming Flexibility, 4
programming port, 44
Programming repeaters into a BaoFeng, 37
Programming simplex channels, 35
Programming simplex channels into a BaoFeng, 35
programming techniques, 154
Propagation Characteristics, 127
Proper Installation, 27
Proper Storage, 144
Protect from Heat and Sun, 5
protecting crucial information, 42
providing a more individualized approach to the radio setting., 21
PTT, 7, 23, 38, 40, 62, 63, 67, 72, 88
public safety radio, 17
pushing a button, 1
Push-to-Talk, 67
Push-To-Talk, 7, 23

Q

Qualified Service, 5
quick and clear communication, 15

R

RADIO ACCESSORIES, 104
RADIO ACCESSORIES AND ADD-ONS, 104
radio antennas, 19, 26
Radio Body, 19
Radio Body and Antenna, 19
Radio Case or Holster, 132
radio communication, 18, 56, 68, 94, 141
Radio Compatibility, 48
radio contact, 43, 47, 55, 67, 94, 126
RADIO ETIQUETTE, 67
RADIO ETIQUETTE AND COMMUNICATION PROTOCOLS, 67
radio experience, 41
radio frequencies, 18, 46, 51, 52, 77, 94, 97
Radio Identification, 45
Radio language and terms, 67

Radio League, 93
Radio Power, 48
Radio Protocol, 68
Radio Regulations, 138, 139, 140, 141
Radio Service, 80, 99
Radio Spectrum Management, 51
Radio to Radio Cloning, 61
radiocommunication services, 138
Radioddity GD-77, 82
radio's range better, 26, 153
Radtel Foldable Tactical Antenna, 19, 26
Range of frequencies, 103
Read and Save Existing Configuration, 45
Read from Radio, 40, 45, 49, 62, 63
Read from the radio, 47
Read the documents, 55
Read the radio's configuration, 58
Reading information, 40
real-time information, 51, 54
Reasonable prices, 4
Receive Frequency, 40
Receiver CTCSS, 40
Receiver CTCSS or DCS, 40
Recreational Use, 14
Reflector gain, 119
Reflector Gain, 119
Regarding privacy and security, 52
Regional and National Allocations, 137
registered repeaters, 21
Registration Number, 18, 99
Regular Audits, 57
Regulations and Legalities, 94
Regulatory Adaptation, 141
regulatory authorities, 52
Regulatory Awareness, 142
Regulatory Compliance and Certification, 142
relevance of adhering to the local legislation, 154
reliable, 5, 8, 9, 12, 13, 49, 113, 130, 153
reliable choice, 9
Remote Areas and remote Communication, 15
Repeat Function, 90
Repeater Operation, 68
Repeating for Best Results, 80
Replacement, 145
Reporting the weather, 97
reprogramme your radio, 96
Requirements, 91, 127
Researching and Identifying Frequencies, 51

Reset/ "Zero Out" The Radio, 74
Respecting Security, 95
Reverse function, 87
RF, 102, 103
right permission or license, 18
Robust Construction, 5
Roger Beep., 92
Ruggedization or waterproofing, 136

S

Safe transmission, 6
Safeguarding Your Communication, 55
Safeguarding Your Radio's Digital DNA, 48
Safety and Emergency Services, 52
Safety from spying, 42
Safety Guidelines, 5
Safety in Dangerous Areas, 5
Safety in Explosive Areas, 5
Safety when charging, 5
SAR, 103
satellite communications, 137, 138
Satellite Orbit Management, 139
satellite paths, 138, 139
Save Configuration, 56
Saves time and effort, 46
Saving and Naming, 63
Saving and Naming Channels, 63
Saving the Configuration, 49
Scan feature, 22
Scan_Add, 40
Scanner and Spectrum Analyzer Tools, 51
Scanning, 5, 71, 85, 86
scanning capabilities, 14
Scanning Channels and Setting Priorities, 71
scanning mode, 40, 85
Scenario-Based Profiles, 56
science studies, 103
scientists, 15
Screen Guard, 132
Search and Rescue Operations, 15
Search for active frequencies and transmissions, 77
Search Operation, 86
Securing the Battery, 28
security activities, 15
Security Blanket for Data Loss, 49
select GPS Info, 96
Select your preferred language, 74

Selecting Channels, 33
Selecting Channels and Frequencies, 33
Selective Calling, 71
sending data, 7, 40, 46, 93
sending data over longer distances, 7
Separation and Isolation, 92
Service Settings, 80
set blocked frequencies, 53
set single frequencies, 35
Setting the Frequencies, 46
Setting up, 40, 58, 96, 112, 146
Setting Up and Managing Radio Groups, 70
Setting up Channels, 40
Seven Steps to Avoid Most Baofeng Radio Problems, 147
Seventy hertz, 78
share data, 15
short-distance contact, 9
short-range conversations, 8
SigCode, 40
Signal Code, 40
Signal interference considerations, 127
Signal penetration through obstacles, 127
signal strength, 13, 51, 117, 122, 124, 126, 127, 130, 131
Signal strength, 130
signaling, 24, 86
Significance and Impact, 140
Simple Interface, 134
Simple Operation, 9
Simple Sharing, 56
simple style and menu options., 41
Simplex and Duplex, 67
Single-Antenna Method, 121
situations or natural disasters, 14
Six Way Charger, 19, 28
six-cell nickel-metal, 19
Size, 114, 133, 134
Slide Down to Release, 30
SMA female connector, 107
SMA-F Female connector, 112
Smooth Connections, 46
soap solution, 143
software bugs, 49
Software called System Fusion, 93
some things to think about, 94, 133
Some types of Baofeng Radios, 6
Specialized Systems, 95

Specific Absorption Rate, 103
specific reasons, 17
Spectrum Allocation and Management, 138
Spectrum Scarcity, 140
Squelch, 9, 32, 67, 68, 69, 78, 80, 86
Squelch Levels, 9
standard antenna, 19
standard Baofeng design, 13
Standing Wave Ratio, 27
Star Button, 24
Status LED, 24
Stay away from extreme temperatures, 145
step-by-step guide for cloning radios, 62
stops searching, 85
Storage Tips, 31
Store the battery correctly, 145
strong features, 13
SUBMIT, 99, 101
successful communication, 154
suitable cord, 46
Summary, 17
Support for Digital Mobile Radio, 4
Support for multiple languages, 134
supporting peaceful living, 139
SWR, 27

T

Tactile marks, 133
Taking out the battery, 29
T-CTCS or T-DCS, 33
TDR, 75, 88, 148
Technological Innovation, 140
technological issue, 42
technology, 3, 4, 43, 55, 97, 138, 152
Telemetry and Sensor Data, 97
Temperature Consideration, 144
Test Function, 80
Testing, 54, 71, 96, 145
Testing and Making Changes, 71
The "Call" button sets, 23
the "Squelch" button, 32
The A/B button, 21, 36
The antenna, 27, 57, 107, 108, 112, 116, 120, 143
The ARRL band plans, 85
The audio cable's 2.5 mm, 91
The Band button, 22
The Baofeng computer tools, 39

The Baofeng PC program, 65
The Baofeng Radio, 1
The Baofeng SRH805S SMA-F, 112
the BAOFENG UV5R Plus UV5RA Plus UV3R Plus, 111
the Baofeng UV-82, 147
the battery level indicator, 24
the BF-F8HP, 13, 17
The BF-F8HP, 13, 14
the BTECH name, 105
The buttons and knobs, 7
the CH-5-6 Gang Charger, 28
The CH-5-6 Gang Charger, 19
The Crucial Role of a Compatible Programming Cable, 46
The design of the GT-3TP, 12
the DM-5R, 3
the EDIT button, 101
the FTDI cable and radio software, 149
the GT-3TP, 12, 17
The Importance of Creating a Backup, 49
The International Telecommunication Union, 137, 138
The LCD screen, 7
The LED on the radio, 40
The Main Display, 20
the Memory option, 63, 70
The Nagoya NA-701C, 26
The Nagoya NA-771, 105, 112
The Nagoya NA-771 15.6-inch whip, 105
the on-screen steps to run CHIRP, 44
The Optional Features window, 63
the radio's menu options, 32
The radio's status LED, 62, 63
the Read button, 62, 63
the software's menu bar, 65
the two memory slots., 65
the USB port, 44
the UV-5R, 3, 7, 17, 74, 75, 77, 78, 81, 82, 84, 106, 109, 148
The UV-5r, 82
The UV-5R is a cheap option, 148
the UV-5R series, 7, 17
The UV-5R Series, 19, 28
The UV-5R series comes equipped with a multitude of features, 7
the UV-5R series stand out, 7
the UV-82 line, 148
the UV-82 series, 11
The VFO/MR button, 21

The VOX, 4
the world of Baofeng radio, 26, 41
Then set the frequency of the offset, 37
Things to Consider Before Buying the Best Antenna for Baofeng, 114
Think about getting an extra speaker-microphone, 57
Time Operation, 85
Tips for Optimizing Baofeng Radio Performance, 57
Tonal frequencies, 78
Tone-burst (1750Hz), 71
Top list of Best Antenna, 105
Top list of Best Antenna for Top 10 model Baofeng Radio, 105
transmission and reception, 67
Transmit Frequency, 40
Transmitter CTCSS, 40
Transmitter CTCSS or DCS., 40
transportation, 15
Transportation, 144
Troubleshooting, 48, 54, 145, 147
Troubleshooting Common Issues, 145, 147
Troubleshooting Issues, 54
Troubleshooting Issues during Testing, 54
Troubleshooting Tips, 48
Troubleshooting Tips for Common Connection Issues, 48
Tune Step column, 82
turn off the radio, 5, 29, 31
Turn off the radio, 5, 39
turn on a Baofeng radio, 31
Turn on the radio, 39
Turning off the Radio, 28
Twin Dual Receive, 148
Two-Antenna Comparison, 121
two-way radios close, 16
TX Power, 40
Type Approval and Certification, 142
type of radio works, 16
Types of Antenna Gain, 117
TYT and Wouxun, 148

U

UHF, 4, 7, 8, 10, 12, 14, 16, 17, 22, 77, 80, 83, 85, 89, 90, 92, 105, 108, 110, 111, 112, 128, 148, 150
UHF radios, 16, 17
Ultra High Frequency, 4, 7, 10, 16, 17, 77, 89, 128
Unauthorized Access, 43

Unboxing and Initial Setup, 26
understand frequencies and channels, 67
understand why antenna gain is so important, 116
Understanding Antenna Gain, 116, 117
UNDERSTANDING ANTENNA GAIN, 116
UNDERSTANDING ANTENNA GAIN AND
   POLARIZATION, 116
Understanding Antenna Polarization, 123
Understanding Baofeng Radios and CHIRP, 41
Understanding Digital Modes, 93
understanding of the programming, 154
Understanding Radio Frequencies, 16
Understanding Radio Language and Terminology, 67
Update as Needed, 57
Updates for the software, 55
Upgraded Antenna, 14
Upload Changes to Radio, 80
Upload to Radio, 46, 48, 53, 61
Uploading to Radio, 46
Use Approved Accessories, 5
Use compatible chargers, 144
use networks in the night, 22
Use the arrow keys, 33, 37, 64, 65
Use the right accessories, 144
Use the Right Accessories, 94
Use the tracking and prioritization options, 71
Use the up and down button keys, 33, 63
Use the UV-5R as an FM radio, 75
useful features and functions, 5
User Education and Awareness, 142
User Interface and Controls, 134
User Manual and Guide, 133
User-Friendly Interface, 5
users flexible contact options, 4
using a Single Antenna, 92
Using and Installing the Antenna, 26
using Baofeng radios, 1, 18
using computer software,, 10
using the keyboard, 21, 38
utilize digital modes on a Baofeng radio, 94
UV-5R accessories, 148
UV-5R file, 82

V

valid amateur radio license, 18
variety of accessories, 4, 13
variety of environments, 154

variety of settings, 7, 9
variety of settings and frequencies, 7
variety of situations, 11, 67, 93, 141, 154
variety of tasks., 98
Velcro strap, 109
Velcro strap and rubber washers, 109
Verification, 50
vertical polarization, 125, 126
Vertical Polarization, 126
Very High Frequency, 4, 7, 10, 16, 17, 77, 89, 128
VFO receivers, 148
VHF radios, 16
Voice Operated Exchange, 7
Voice Prompts, 134
Voice-Activated Transmission, 4, 9
Voice-Operated Exchange, 11
Volunteer Groups, 15
VOX, 7, 9, 11, 13, 14, 91

W

Walkie Talkie Antenna, 105, 112
walkie-talkie, 1, 72, 152
Walking through the Connection Process, 47
Water-Resistant Bag or Cover, 133
Wavelength relative to object size, 127
WB4APR, 97
weaker signals, 32
Weather-Appropriate Gear, 133
well-known options, 26
What Are the Dangers of Open Frequencies?, 43
What is a Baofeng?, 151
What is meant by "programming", 35
Which Baofeng radio is the best?, 153
Why is antenna polarization important, 125
Why keeping track of the configuration is a good idea,
   56
Why Should You Invest in Baofeng Antenna?, 104
wide range of communication needs., 6, 8, 14
wide range of frequencies, 7, 10, 12, 17, 33
Wideband, 40
Wideband or Narrowband operation, 40
wider broadcast area, 16
Wi-Fi antennas, 26
Windows, 44, 61
wireless conversations, 17
Wiring and Connections, 96
work together in rural places and outdoor settings., 15

working in cold places, 31

works with your Baofeng radio, 39, 96

world of radio communication, 53

World Radiocommunication Conferences, 139

WRC, 139

WRCs, 139

Write Data to Radio window, 63

Write the modified configuration to the radio, 58

Y

Yagi-Uda antenna, 120

You might be wondering why I didn't tell you to add any tones., 84

Your Path to Radio Programming, 43

www.ingramcontent.com/pod-product-compliance
Lightning Source LLC
Chambersburg PA
CBHW082211290526
45794CB00009B/3500